# Lecture Notes in Physics

## New Series m: Monographs

**Managing Editor**

W. Beiglböck
Assisted by Mrs. Sabine Landgraf
c/o Springer-Verlag, Physics Editorial Department II
Tiergartenstrasse 17, D-69121 Heidelberg, Germany

# The Editorial Policy for Monographs

The series Lecture Notes in Physics reports new developments in physical research and teaching - quickly, informally, and at a high level. The type of material considered for publication in the New Series m includes monographs presenting original research or new angles in a classical field. The timeliness of a manuscript is more important than its form, which may be preliminary or tentative. Manuscripts should be reasonably self-contained. They will often present not only results of the author(s) but also related work by other people and will provide sufficient motivation, examples, and applications.

The manuscripts or a detailed description thereof should be submitted either to one of the series editors or to the managing editor. The proposal is then carefully refereed. A final decision concerning publication can often only be made on the basis of the complete manuscript, but otherwise the editors will try to make a preliminary decision as definite as they can on the basis of the available information.

Manuscripts should be no less than 100 and preferably no more than 400 pages in length. Final manuscripts should preferably be in English, or possibly in French or German. They should include a table of contents and an informative introduction accessible also to readers not particularly familiar with the topic treated. Authors are free to use the material in other publications. However, if extensive use is made elsewhere, the publisher should be informed. Authors receive jointly 50 complimentary copies of their book. They are entitled to purchase further copies of their book at a reduced rate. As a rule no reprints of individual contributions can be supplied. No royalty is paid on Lecture Notes in Physics volumes. Commitment to publish is made by letter of interest rather than by signing a formal contract. Springer-Verlag secures the copyright for each volume.

# The Production Process

The books are hardbound, and quality paper appropriate to the needs of the author(s) is used. Publication time is about ten weeks. More than twenty years of experience guarantee authors the best possible service. To reach the goal of rapid publication at a low price the technique of photographic reproduction from a camera-ready manuscript was chosen. This process shifts the main responsibility for the technical quality considerably from the publisher to the author. We therefore urge all authors to observe very carefully our guidelines for the preparation of camera-ready manuscripts, which we will supply on request. This applies especially to the quality of figures and halftones submitted for publication. Figures should be submitted as originals or glossy prints, as very often Xerox copies are not suitable for reproduction. For the same reason, any writing within figures should not be smaller than 2.5 mm. It might be useful to look at some of the volumes already published or, especially if some atypical text is planned, to write to the Physics Editorial Department of Springer-Verlag direct. This avoids mistakes and time-consuming correspondence during the production period.

As a special service, we offer free of charge $\text{L\!A\!T\!}_E\text{X}$ and $\text{T}_E\text{X}$ macro packages to format the text according to Springer-Verlag's quality requirements. We strongly recommend authors to make use of this offer, as the result will be a book of considerably improved technical quality.

Manuscripts not meeting the technical standard of the series will have to be returned for improvement.

For further information please contact Springer-Verlag, Physics Editorial Department II, Tiergartenstrasse 17, D-69121 Heidelberg, Germany.

Olivier Piguet   Silvio P. Sorella

# Algebraic Renormalization

## Perturbative Renormalization, Symmetries and Anomalies

 Springer

**Authors**

Olivier Piguet
Département de Physique Théorique
Université de Genève
CH-1211 Genève 4, Switzerland

Silvio P. Sorella
Centro Brasileiro de Pesquisas Fisicas
Rua Xavier Sigaud 150, 22290-180 Urca
Rio de Janeiro, Brazil

ISBN 978-3-662-14040-6    ISBN 978-3-540-49192-7 (eBook)
DOI 10.1007/978-3-540-49192-7

CIP data applied for.

Originally published by Springer-Verlag Berlin Heidelberg New York in 1995
Softcover reprint of the hardcover 1st edition 1995

Typesetting: Camera-ready by the author
SPIN: 10127325    55/3142-543210 - Printed on acid-free paper

# Preface

The idea of this book originated from two series of lectures given by us at the Physics Department of the Catholic University of Petrópolis, in Brazil. Its aim is to present an introduction to the "algebraic" method in the perturbative renormalization of relativistic quantum field theory. Although this approach goes back to the pioneering works of Symanzik in the early 1970s and was systematized by Becchi, Rouet and Stora as early as 1972–1974, its full value has not yet been widely appreciated by the practitioners of quantum field theory. Becchi, Rouet and Stora have, however, shown it to be a powerful tool for proving the renormalizability of theories with (broken) symmetries and of gauge theories. We have thus found it pertinent to collect in a self–contained manner the available information on algebraic renormalization, which was previously scattered in many original papers and in a few older review articles.

Although we have taken care to adapt the level of this book to that of a post-graduate (Ph.D.) course, more advanced researchers will also certainly find it useful.

The deeper knowledge of renormalization theory we hope readers will acquire should help them to face the difficult problems of quantum field theory. It should also be very helpful to the more phenomenology oriented readers who want to familiarize themselves with the formalism of renormalization theory, a necessity in view of the sophisticated perturbative calculations currently being done, in particular in the standard model of particle interactions.

*Acknowledgements.* We particularly thank Profs. Renato M. Doria and José Helayel Neto for their very kind invitations to Petrópolis and the students who had the patience to hear our courses and encouraged us by their pertinent questions.

We wish to express our gratitude to C. Becchi, A. Blasi, R. Collina, F. Delduc, R.M. Doria, S. Emery, F.· Gieres, E. Guadagnini, J. Helayel Neto, C. Lucchesi, N. Maggiore, F.A.B. Rabelo de Carvalho, M. Schweda, K. Sibold, C. Stafford, R. Stora, M. Werneck de Oliveira, J. Wess and S. Wolf, as well as to all our colleagues in the physics departments of the Universities of Geneva and Vienna, and at the Centro Brasileiro Pesquisas Físicas (CBPF) in Rio de Janeiro, some of them for fruitful collaborations and all of them for many helpful discussions on the various aspects of the material presented here.

This book was made possible thanks to the generous financial support of the Conselho Nacional de Pesquisa e Desenvolvimento (CNPq, Brazil), the Austrian Science Foundation, the Ausseninstitut of the Technical University of Vienna and the Swiss National Science Foundation.

*Geneva and Rio de Janeiro,*
*February 1995*                                        *Olivier Piguet and Silvio P. Sorella*

# Table of Contents

# 1. Introduction

Relativistic quantum field theory (RQFT) was introduced at the end of the 1920s in order to unify the principles of quantum mechanics and of special relativity (see for example the first of Dirac's papers in [1]). Since that time RQFT has been the object of ever increasing interest. Together with some of its extensions, such as string theory, it has now become our main conceptual framework for describing the fundamental structure of matter [2]. We underline, in particular, the well established and successful applications to the gauge theories of particle interactions: QED, the standard model, QCD and supersymmetric extensions thereof. Although a quantum theory of gravity has not yet been firmly formulated, there are now hopes for a consistent field theoretical description as well [3].

However, it has been realized since its beginnings (see [4] and Heisenberg's paper in [1]) that RQFT was plagued by an apparently insuperable difficulty, namely the problem of the ultraviolet (UV) divergences, which had forbidden any computation beyond the lowest order of perturbation theory. The origin of this difficulty lies in the occurrence of singular products of distributions at coinciding points in the computation of the higher–order terms [5, 6]. As is now well known, the search for a solution to the problem of the subtraction of the UV divergences, lasting from the 1940s to the 1970s, has lead to the establishment of a consistent theoretical and mathematical framework called renormalization theory [1, 7, 8]. These efforts, rewarded first by the spectacular agreement of theory with experiment in QED [1, 9], culminated in the proof of the renormalizability of the nonabelian gauge theories [10, 11], which are the basis of our present understanding of high–energy physics.

At the beginning of the 1970s a few UV subtraction schemes were available and were proven to be equivalent [7]. It is a remarkable result that these schemes led to a renormalized perturbation theory whose properties can be collected in a set of rigorous statements, common to all of them. These statements, based on locality and power counting, constitute the content of the so–called quantum action principle (QAP) [12, 13, 14]. The latter is the starting point of any rigorous investigation in the framework of perturbation theory.

It was early recognized by Becchi, Rouet and Stora, in their works on the BRS invariance of gauge theories [15, 16] and on the models with broken rigid symmetries [17], that the use of the QAP leads to the possibility of a fully algebraic proof of the renormalizability of a theory characterized by a set of local or rigid invariances. Let us remark that we distinguish "renormalization", which for us means the implementation of the symmetries of a given classical model to all orders of perturbation series, from the mere "UV subtractions". The QAP actually allows one to control the breaking of a symmetry induced by a noninvariant subtraction scheme, helping then to give an algebraic answer about the possibility of restoring the symmetry through the addition of compensating noninvariant local counterterms. It is worthwhile to emphasize that such algebraic proofs do not rely on the existence of a regularization preserving the symmetries.

The formulation of renormalization theory presented in this book concerns only the perturbative expansion in the powers of some appropriate parameter. We shall not touch on any nonperturbative aspect of RQFT. Moreover, "perturbation theory" will mean here that the quantities of interest are computed order by order, without any attempt to give a meaning to the sum. Indeed, except for a few lower–dimensional examples [18], there is no rigorous result about the convergence properties of the perturbation series. However, the remarkable agreement between experiment and per-turbative theoretical calculations in QED [19], in the electroweak model [20] as well as in high energy QCD [21] thanks to asymptotic freedom, suggests that perturbation theory is a reliable approximation scheme in the regime of weak couplings.

The aim of this book is to provide a pedagogical and self–contained introduction to the algebraic method of renormalization. This will be done by considering several examples from rigid symmetries and gauge theories to topological models, and also to string theory viewed as a two–dimensional field theory. The choice of the examples is intended to give a rather complete sample of the domain of applicability of algebraic renormalization. Let us underline that our point of view here is basically "Lagrangian", based on the QAP, as opposed to that of the canonical quantization approach [22, 23].

We begin with an introduction to the ultraviolet problem and to its solution by means of a subtraction procedure, chosen for illustration purposes to be the mo-mentum space BPHZ method [24, 25]. We then proceed with the statement of the quantum action principle and its use in simple models with rigid symmetries. As a natural continuation, the renormalizability of gauge theories of the Yang–Mills type is proven. A special chapter is then devoted to the basic aspects of BRS cohomology, systematically used throughout the book in order to characterize the renormalization effects, i.e., the counterterms and the anomalies. Among the applications, one will find a detailed presentation of the nonrenormalization theorems of the chiral anomalies, examples of ultraviolet finite topological models, and the field–theory treatment of the bosonic string.

We hope finally to convince the reader of the usefulness of the method of algebraic renormalization, which is particularly adapted to the study of theories containing chiral fermions or the Levi–Civita tensor and, more generally, to any theory for which no invariant regularization is known, such as supersymmetric gauge theories [26], for example.

# 2. Renormalized Perturbation Theory

In this chapter we shall recall some basic facts of renormalization theory. The proofs will be omitted.

The objects of interest are the Green functions, i.e., the vacuum expectation values of time–ordered products of field operators. Indeed all relevant physical quantities, such as the matrix elements of the scattering operator and of the field operators, can be computed from them [27]. The existence and the properties of the Green functions will be studied in perturbation theory, i.e., as series in powers of the coupling constant(s) or, equivalently as we shall see, in powers of Planck's constant. These expansions will be treated as *formal power series*. This means that we shall be interested only in their coefficients and not at all in their sums. Hence all the identities we are going to meet through this book will hold order by order. Two elementary but nevertheless very important facts about formal power series which we shall systematically exploit are the following:

1. A formal power series vanishes if all its terms vanish. Equivalently, it is nonvanishing if and only if at least one of its terms is nonvanishing.

2. A formal power series admits an inverse which is also a power series if and only if its zeroth–order term is invertible. For example, if the generic term is an $N \times N$ matrix, the zeroth–order term must be an invertible $N \times N$ matrix.

The results stated in this chapter are quite general. However, in order to keep the presentation as simple as possible, all explicit examples and computations are done for the case of the $\varphi_4^4$ theory, i.e., the theory of a scalar field $\varphi$ in four–dimensional Euclidean space–time, with a self–interaction $\varphi^4$. This theory is given by the action[1]

$$ S = \int d^4x \, \mathcal{L}(x) = \int d^4x \, \left( \frac{1}{2} \partial^\mu \varphi \partial_\mu \varphi + \frac{m^2}{2} \varphi^2 + \frac{g}{4!} \varphi^4 \right) . \tag{2.1}$$

The free causal propagator is

$$\Delta(x - y) = \frac{1}{(2\pi)^4} \int d^4k \, e^{-ik(x-y)} \tilde{\Delta}(k) ,$$
$$\tilde{\Delta}(k) = \frac{\hbar}{k^2 + m^2} . \tag{2.2}$$

---

[1]**Notations and Conventions:** The units are chosen such that

$$c = \hbar = 1 .$$

We shall, however, sometimes keep Planck's constant $\hbar$ explicit since we are going to consider perturbation theory as a formal power series in $\hbar$.

If not otherwise stated, the space–time metric will be the $D$–dimensional flat Euclidean one. It is known [28] that under suitable assumptions, which are realized in perturbation theory, the existence of the Euclidean theory guaranties the existence of a corresponding theory in Minkowski space–time. The Minkowskian theory is obtained formally from the Euclidean one by an analytic continuation in the complex energy plane, which amounts to replacing the 0th component $p_0$ of the four–momentum by $-ip_0$. This is called the Wick rotation.

*Remark.* In Minkowski space–time the Lagrangian and the free causal propagator of the $\varphi_4^4$ theory read

$$\mathcal{L} = \frac{1}{2}\partial^\mu\varphi\partial_\mu\varphi - \frac{m^2}{2}\varphi^2 - \frac{g}{4!}\varphi^4 \ ,$$

$$\tilde{\Delta}(k) = \lim_{\varepsilon\to +0} \frac{i\hbar}{k^2 - m^2 + i\varepsilon} \ .$$

The $\varepsilon$ prescription above follows from the definition we have given to the Wick rotation. It corresponds to the causal prescription of Stueckelberg and Feynman [29, 30].

## 2.1 Generating Functionals

Let us consider a theory involving a set of fields $\phi_i(x)$ in $D$–dimensional space–time[2], with the index $i$ denoting the species as well as the spin and internal degrees of freedom. The (classical) dynamics is defined by the action

$$S(\phi) = \int dx \mathcal{L}(x) = S_0(\phi) + S_{\text{int}}(\phi) \ . \tag{2.3}$$

The Lagrangian has the general form

$$\mathcal{L}(x) = \frac{1}{2}\phi_i(x)K^{ij}(\partial)\phi_j(x) + \mathcal{L}_{\text{int}} = \mathcal{L}_0 + \mathcal{L}_{\text{int}} \ . \tag{2.4}$$

$K^{ij}(\partial)$ is some invertible differential operator, usually a polynomial of second order in $\partial$ for the bosonic fields and of first order for the fermionic ones: this quadratic piece of the Lagrangian corresponds to the free theory whereas $\mathcal{L}_{\text{int}}$ describes the interactions.

### 2.1.1 The Green Functional

As we have said, the objects of the corresponding quantum theory one wants to compute are the Green functions, i.e., the vacuum expectation values of the time–ordered products of field operators:

$$G_{i_1\cdots i_N}(x_1, \cdots, x_N) = \langle T\phi_{i_1}(x_1)\cdots\phi_{i_N}(x_N)\rangle \ . \tag{2.5}$$

These Green functions may be collected together in the generating functional $Z(J)$, a formal power series in the "classical sources" $J^i(x)$:

---

[2]Space–time point coordinates are denoted by $(x^\mu, \ \mu = 0, \cdots, D-1)$.

$$Z(J) = \sum_{N=0}^{\infty} \frac{(-1/\hbar)^N}{N!} \int dx_1 \cdots dx_N \, J^{i_1}(x_1) \cdots J^{i_N}(x_N) \, G_{i_1 \cdots i_N}(x_1, \cdots, x_N) \, . \quad (2.6)$$

The Green functions are tempered distributions. The sources $J(x)$ thus belong to the set of Schwartz fast decreasing $C^\infty$ functions ("test functions").

The Green functional (2.6) is *formally* given by the Feynman path integral

$$Z(J) = \mathcal{N} \int \mathcal{D}\phi \, e^{-\frac{1}{\hbar}\left(S(\phi) + \int dx J^i(x)\phi_i(x)\right)} \, , \quad (2.7)$$

where $\mathcal{N}$ is some (generally ill–defined) numerical factor. The solution for the free theory ($\mathcal{L}_{\text{int}} = 0$ in (2.4)) is given by

$$Z_{\text{free}}(J) = e^{\frac{1}{2\hbar^2} \int dx_1 dx_2 J^{i_1}(x_1) J^{i_2}(x_2) \Delta_{i_1 i_2}(x_1, x_2)} \, , \quad (2.8)$$

where $\Delta_{i_1 i_2}(x_1, x_2)$ is the Stueckelberg–Feynman free causal propagator, obtained by inverting[3] the wave operator $K^{ij}(\partial)$ of (2.4):

$$K^{ij}\Delta_{jk}(x, y) = \hbar \delta_k^i \delta^D(x - y) \, . \quad (2.9)$$

In the case of the full interacting theory a formal solution is given by [31]

$$Z(J) = \mathcal{N} e^{-\frac{1}{\hbar} S_{\text{int}}\left(-\hbar \frac{\delta}{\delta J}\right)} Z_{\text{free}}(J) \, . \quad (2.10)$$

This expression leads to the well–known perturbative expansion of the Green functions in terms of Feynman graphs.

**Example 2.1 The two–point function of the $\varphi_4^4$ theory.**
The expansion of $Z_{\text{free}}$ reads

$$\begin{aligned}
Z_{\text{free}}(J) = \; & 1 + \frac{1}{2\hbar^2} J^{i_1} J^{i_2} \Delta_{i_1 i_2} + \frac{1}{2!4\hbar^4} J^{i_1} J^{i_2} J^{i_3} J^{i_4} \Delta_{i_1 i_2} \Delta_{i_3 i_4} \\
& + \frac{1}{3!8\hbar^6} J^{i_1} J^{i_2} J^{i_3} J^{i_4} J^{i_5} J^{i_6} \Delta_{i_1 i_2} \Delta_{i_3 i_4} \Delta_{i_5 i_6} + \cdots \, .
\end{aligned} \quad (2.11)$$

The indices now represent both the discrete indices (none in the present example) and the space–time points. Repeated indices mean both summation and integration. The two–point Green function is given by ($\delta_p$ stands for $\delta/\delta J^p$)

$$\begin{aligned}
\hbar^2 \, \delta_1 \delta_2 Z(J)\big|_{J=0} &= \mathcal{N}\hbar^2 \, \delta_1 \delta_2 e^{-\frac{g\hbar^4}{4!}(\delta_k)^4} Z_{\text{free}}(J)\Big|_{J=0} \\
&= \mathcal{N}\hbar^2 \delta_1 \delta_2 \left( Z_{\text{free}}(J) - \frac{g\hbar^3}{4!}(\delta_k)^4 Z_{\text{free}}(J) \right)\Big|_{J=0} + \cdots \\
&= \Delta_{12} - \frac{g\hbar^5}{4!} \frac{1}{3!8\hbar^6} \delta_1 \delta_2 \delta_k \delta_k \delta_k \delta_k J^{i_1} J^{i_2} J^{i_3} J^{i_4} J^{i_5} J^{i_6} \Delta_{i_1 i_2} \Delta_{i_3 i_4} \Delta_{i_5 i_6} \\
&= \Delta_{12} - \frac{g}{2\hbar} \Delta_{1i} \Delta_{ii} \Delta_{i2} + \cdots \, .
\end{aligned} \quad (2.12)$$

(Disconnected terms containing vacuum graph factors, such as $\Delta_{12}\Delta_{ii}\Delta_{ii}$, are absorbed in the factor $\mathcal{N}$.)

---

[3] Signs in the following equations correspond to the case where the fields $\phi_i$ are all bosonic. The reader may generalize to the case where fermionic fields are also present.

One sees explicitly here that the expansion in $g$ is also a loop expansion and corresponds to an expansion in $\hbar$ – a general argument will be given in Subsect. 2.1.3. Recalling that the free propagator is proportional to $\hbar$ (cf. (2.2)), one indeed sees that the first term is of order $\hbar$ and the second one of order $\hbar^2$. The drawing of the corresponding Feynman graphs is left to the reader. He will see in particular that the factor $\Delta_{ii}$ in the second term of the last line of (2.12) represents a loop made of one propagator with coinciding end points. This object being the "value" of the propagator (2.2) at $x - y = 0$, and the propagator being a distribution, this "value" may be expected to be ill defined, and is in fact not defined! $\Delta_{ii}$ is indeed equal to the singular momentum–space integral

$$\Delta(0) = \frac{1}{(2\pi)^4} \int d^4k \, \frac{\hbar}{k^2 + m^2} \, . \tag{2.13}$$

It is the aim of renormalization theory to give a meaning to such ill–defined expressions. This will be explained in Sect. 2.2.

### 2.1.2 The Connected and the Vertex Functionals

Let us introduce two more functionals. The contributions of all *connected* graphs to a Green function is called a truncated or a connected Green function. The generating functional of the connected Green functions

$$Z^c(J) =$$
$$\sum_{N=1}^{\infty} \frac{(-1/\hbar)^{N-1}}{N!} \int dx_1 \cdots dx_N \, J^{i_1}(x_1) \cdots J^{i_N}(x_N) \, \langle T\phi_{i_1}(x_1) \cdots \phi_{i_N}(x_N) \rangle_{\text{conn}}$$
$$\tag{2.14}$$

is related to the Green functional $Z(J)$ by

$$Z(J) = e^{-\frac{1}{\hbar} Z^c(J)} \, . \tag{2.15}$$

A *one–particle irreducible* (1PI) graph is a connected graph, amputated from its external legs, which remains connected after cutting any internal line. The contribution of these graphs to an (amputated) connected Green function is called a 1PI or vertex function. The generating functional of the vertex functions reads

$$\Gamma(\phi^{\text{class}}) = \sum_{N=2}^{\infty} \frac{1}{N!} \int dx_1 \cdots dx_N \, \phi_{i_1}^{\text{class}}(x_1) \cdots \phi_{i_N}^{\text{class}}(x_N) \, \Gamma^{i_1 \cdots i_N}(x_1, \cdots, x_N) \tag{2.16}$$
$$\Gamma^{i_1 \cdots i_N}(x_1, \cdots x_N) = \langle T\phi_{i_1}(x_1) \cdots \phi_{i_N}(x_N) \rangle_{\text{1PI}} \, ,$$

where the arguments $\phi^{\text{class}}$, the "classical fields", are Schwartz fast decreasing test functions. Later on we shall suppress the superscript "class", no confusion between the classical field and the corresponding quantum field being expected. The vertex functional is related to the connected functional by a Legendre transformation:

$$\Gamma(\phi) = Z^c(J) - \int dx J^i(x)\phi_i(x) \Big|_{\phi_i(x) = \frac{\delta Z^c}{\delta J^i(x)}} \, . \tag{2.17}$$

(In the right–hand side, $J(x)$ is replaced by the solution $J(\phi)(x)$ of the equation $\phi_i(x) = \delta Z^c/\delta J^i(x)$.) The inverse Legendre transformation is given by

$$Z^c(J) = \Gamma(\phi) + \int dx J^i(x)\phi_i(x)\Big|_{J^i(x)=-\frac{\delta\Gamma}{\delta\phi_i(x)}} . \tag{2.18}$$

We have assumed that the vacuum expectation values of the field variables are zero. One has thus

$$\frac{\delta Z^c}{\delta J^i(x)}\Big|_{J=0} = 0 , \qquad \frac{\delta\Gamma}{\delta\phi_i(x)}\Big|_{\phi=0} = 0 . \tag{2.19}$$

*Remark.* For the two–points functions the Legendre transformation yields:

$$\int dy \Gamma^{ij}(x,y)\,\langle T\phi_j(y)\phi_k(z)\rangle_{\text{conn}} = -\delta^i_k\delta^D(x-y) , \tag{2.20}$$

which is the all order generalization of (2.9).

### 2.1.3  Expansion in $\hbar$

From its definition and the formula (2.10) for the perturbative expansion of the Green functional one can easily check that the vertex functional can be written as a formal power series in $\hbar$:

$$\Gamma(\phi) = \sum_{n=0}^{\infty}\hbar^n\Gamma^{(n)}(\phi) , \tag{2.21}$$

the order $n$ corresponding to the contributions of the $n$–loop graphs. In order to prove this statement, let us consider the contribution of a 1PI diagram consisting of $I$ internal lines, $V$ vertices and $L$ loops. Counting a factor $\hbar$ for each internal line, a factor $\hbar^{-1}$ for each vertex and an overall factor $\hbar$ due to the factor $\hbar^{-1}$ in (2.15), we find the value $I - V + 1$ for the total power in $\hbar$. The result then follows from the topological identity

$$L = I - V + 1 , \tag{2.22}$$

due to Euler. The zeroth order

$$\Gamma^{(0)}(\phi) = S(\phi) \tag{2.23}$$

is the classical action (2.3). This is obvious since the only 1PI zero-loop graphs – the 1PI tree graphs – are the trivial ones, i.e., those containing a single vertex, and this vertex corresponds to a term of the interaction Lagrangian. In this approximation the Legendre transform $Z^{c(0)}(J)$ of $\Gamma^{(0)}(\phi)$ generates the connected Green functions, given by the connected tree Feynman graphs. $\Gamma^{(n)}$ corresponds to the contributions of the $n$–loop graphs.

### 2.1.4  Composite Fields

We are also interested in Green functions involving composite field operators. Such operators appear in particular in theories invariant under field transformations which depend nonlinearly on the fields. Let us thus consider field operators $Q^p(x)$, corresponding to local field polynomials $Q^p_{class}(x)$ in the classical theory. If one performs again the construction above, but starting with the new classical interaction

$$S_{int}(\phi, \rho) = S_{int}(\phi) + \int dx \rho_p(x) Q^p_{class}(x) \tag{2.24}$$

depending on the "external fields" $\rho_p(x)$, one obtains a new Green functional [32]

$$Z(J, \rho) = \sum_{N=0}^{\infty} \sum_{M=0}^{\infty} \frac{(-1/\hbar)^N}{N!} \frac{(-1/\hbar)^M}{M!} \int dx_1 \cdots dx_N \int dy_1 \cdots dy_M$$
$$J^{i_1}(x_1) \cdots J^{i_N}(x_N) \, \rho_{p_1}(y_1) \cdots \rho_{p_M}(y_M) \tag{2.25}$$
$$\langle T\phi_{i_1}(x_1) \cdots \phi_{i_N}(x_N) \, Q^{p_1}(y_1) \cdots Q^{p_M}(y_M) \rangle \ ,$$

which generates all the Green functions with the insertions of the local composite quantum fields $Q^p(x)$.

The connected functional $Z^c(J, \rho)$ and the vertex functional $\Gamma(\phi, \rho)$ involving these composite fields are related to $Z(J, \rho)$ via the generalizations of (2.15), (2.17) and (2.18):

$$Z(J, \rho) = e^{-\frac{1}{\hbar} Z^c(J, \rho)} \ , \tag{2.26}$$

$$\Gamma(\phi, \rho) = Z^c(J, \rho) - \int dx J^i(x)\phi_i(x)\Big|_{\phi_i(x) = \frac{\delta Z^c}{\delta J^i(x)}} \ , \tag{2.27}$$

and

$$Z^c(J, \rho) = \Gamma(\phi, \rho) + \int dx J^i(x)\phi_i(x)\Big|_{J^i(x) = -\frac{\delta \Gamma}{\delta \phi_i(x)}} \ . \tag{2.28}$$

In particular

$$-\hbar \frac{\delta Z}{\delta \rho_p(y)}\Big|_{\rho=0} := Q^p(y) \cdot Z(J) \ , \tag{2.29}$$

generates the Green functions

$$\langle T Q_p(y) \, \phi_{i_1}(x_1) \cdots \phi_{i_N}(x_N) \rangle \ , \tag{2.30}$$

whose Feynman graphs contain a new vertex corresponding to the insertion of the field polynomial $Q^p_{class}(y)$ (with possible quantum corrections). In the same way

$$\frac{\delta Z^c}{\delta \rho_p(y)}\Big|_{\rho=0} = Q^p(y) \cdot Z^c(J) \ , \tag{2.31}$$

and

$$\frac{\delta \Gamma}{\delta \rho_p(y)}\Big|_{\rho=0} = Q^p(y) \cdot \Gamma(\phi) \ , \tag{2.32}$$

generate connected and 1PI Green functions, respectively, with the insertion of the operator $Q^p(y)$. The zeroth order term of the loop ($\hbar$) expansion of (2.32) coincides with the classical field polynomial which is the starting point of the perturbative construction of the quantum insertion $Q^p$:

$$Q^p(y) \cdot \Gamma(\phi) = Q^p_{class}(y) + O(\hbar) \ . \tag{2.33}$$

## 2.2 Renormalization

We shall discuss the renormalization of the Green functional $Z(J, \rho)$ or, preferably although this is equivalent, of the vertex functional $\Gamma(\phi, \rho)$: the latter is indeed the appropriate object to consider in this respect since the vertex functions contain all the loop integrals, hence the divergences. We consider the quantum fields $\phi_i$ and the external fields $\rho_p$ together because we are also interested in the renormalization of the composite fields $Q^p$, which occur for example in nonlinear field transformations. This renormalization will thus automatically be included in the following treatment. The fields $\phi_i$ and $\rho_p$ will be collectively denoted by $\Phi_i$.

### 2.2.1 Ultraviolet Divergences

The ultraviolet (UV) divergences, i.e., the singularities of the momentum space integrals caused by a bad high momentum behaviour of their integrands, affect the one–particle irreducible (1PI) graphs. We shall thus consider here graphs contributing to the Fourier transform $\tilde{\Gamma}$, defined by

$$(2\pi)^D \delta^D (\textstyle\sum p_k) \tilde{\Gamma}_{N_1 N_2 \cdots}(p_1, \cdots, p_{N_1}, p_{N_1+1}, \cdots, p_{N_1+N_2}, \cdots)$$
$$= \int \left( \prod d^D x_k \right) e^{i \sum x_k p_k} \Gamma_{N_1 N_2 \cdots}(x_1, \cdots, x_{N_1}, x_{N_1+1}, \cdots, x_{N_1+N_2}, \cdots) , \tag{2.34}$$

of a generic translation invariant vertex function

$$\Gamma_{N_1 N_2 \cdots}(x_1, \cdots, x_{N_1}, x_{N_1+1}, \cdots, x_{N_1+N_2}, \cdots)$$
$$= \langle T \Phi_1(x_1) \cdots \Phi_1(x_{N_1}) \Phi_2(x_{N_1+1}) \cdots \Phi_2(x_{N_1+N_2}) \cdots \rangle_{1\text{PI}} , \tag{2.35}$$

where $N_A$ denotes the number of times the field species $\Phi_A$ appears.

Let us consider the contribution of a 1PI graph $\gamma$ with:

- $N_A$ external (amputated) legs corresponding to the field $\Phi_A = \phi_i$ or $\rho_p$ of (UV) dimension $d_A$, $A = 1, 2, \cdots$;

- $N_V$ interaction vertices $V_k$ of (UV) dimension $d_k$, $k = 1, \cdots, N_V$ (vertices from the interaction Lagrangian);

- $I_{ij}$ internal lines corresponding to the free propagator $\Delta_{ij}$, $i, j = 1, 2, \cdots$;

- L loops.

The UV dimensions $d_i$ of the quantum fields $\phi_i$ are defined in such a way that

(i) the total UV dimension of each term of the quadratic part of the free Lagrangian is bounded by the space–time dimension $D$, the dimension of the space–time derivative $\partial_\mu$ being taken equal to 1.

(ii) the UV dimensions $d_i$ must obey the inequalities [24]

$$d_i + d_j \geq d_{ij} + D , \tag{2.36}$$

where the UV dimension $d_{ij}$ of the free propagator $\tilde{\Delta}_{ij}(p)$ in momentum space, i.e., of the Fourier transform of

$$\Delta_{ij}(x - y) = \langle T\phi_i(x)\phi_j(y)\rangle_{\text{free}} ,\qquad (2.37)$$

is defined by its asymptotic behaviour

$$\tilde{\Delta}_{ij}(p) \sim p^{d_{ij}} , \quad p \to \infty . \qquad (2.38)$$

To an external field $\rho_p$ coupled to a composite field $Q^p$ of dimension $d_{Q^p}$ one assigns a dimension $d_p = D - d_{Q^p}$. Each vertex of the graph corresponds to a monomial made of the fields $\phi_i$ and $\rho_p$ and of their derivatives. The UV dimension of the vertex is the dimension of this monomial.

To the graph $\gamma$ corresponds (formally) a momentum–space integral

$$J_\gamma(p) = \left[ \int \prod_{l=1}^{L} d^D k_l \right] I_\gamma(p, k) . \qquad (2.39)$$

Here $p = (p_1, \cdots, p_N)$, with $\sum_1^N p_k = 0$, denotes the $N$ external momenta (taken as incoming), and $k = (k_1, \cdots, k_L)$ the $L$ independent internal momenta. Under a scale transformation $p \to \lambda p$, $k \to \lambda k$, the integrand being a rational function of the momenta behaves as a power of $\lambda$ for $\lambda \to \infty$:

$$\left[ \prod_{l=1}^{L} d^D(\lambda k_l) \right] I_\gamma(\lambda p, \lambda k) \sim \lambda^{d(\gamma)} . \qquad (2.40)$$

The power $d(\gamma)$ defines the *superficial UV degree of divergence* of the graph $\gamma$. It is equal to

$$d(\gamma) = D + \sum_{\text{all vertices } V_k} (\dim(V_k) - D) - \sum_A N_A d_A . \qquad (2.41)$$

If all interaction terms in the Lagrangian (2.4) are of dimension less or equal to $D$ one sees that the superficial degree of divergence is bounded by an expression which does not depend on the number of vertices, i.e., which is independent of the order:

$$d(\gamma) \le d_{\max}(\gamma) = D - \sum_A N_A d_A \qquad (2.42)$$

In such a case the theory belongs to the class of the so–called *power–counting renormalizable* theories. We shall say more on this point in Subsect. 2.2.5.

If the diagram $\gamma$ represents a contribution to a vertex function with only one external field $\rho$ of dimension $d_\rho$, i.e., with the insertion of only one composite field operator $Q(x)$ of dimension $d_Q = D - d_\rho$ (cf. (2.30)), then one of its vertices corresponds to the composite field, whereas all other vertices correspond to the terms of the interaction Lagrangian. In this case, and if moreover the theory is power–counting renormalizable, the upper bound for the superficial degree of divergence reads

$$d_{\max}(\gamma) = d_Q - \sum_i N_i d_i , \qquad (2.43)$$

**Example 2.2** In the example of the $\varphi_4^4$ theory (2.1), the superficial degree of divergence of a graph $\gamma$ with $N$ external legs is

$$d(\gamma) = d_{\max}(\gamma) = 4 - N \ . \tag{2.44}$$

If mass counterterms $\varphi^2$ are added to the Lagrangian (2.1) then

$$d(\gamma) = 4 - N - 2N_m \tag{2.45}$$

where $N_m$ is the number of mass insertions in the diagram $\gamma$.

## 2.2.2 Ultraviolet Subtractions

As it is well known the momentum–space integrals associated with the Feynman graphs are generally UV divergent. A theorem due to Weinberg [33] and later on proved for Minkowski space–time by Zimmermann [34] gives a criterium of absolute convergence for such integrals (we don't consider for the moment theories with massless fields):

**Theorem 2.3** *(Power–Counting Theorem) Provided all free propagators have non–zero masses, the multiple integral (2.39) associated to the graph $\gamma$ is absolutely convergent if and only if the superficial degree of divergence of $\gamma$ and of each of its 1PI subgraphs is strictly negative.*

The ultraviolet divergences arise from the illegal attempt to multiply distributions (the free propagators) locally, i.e., with coinciding arguments.

**Example 2.4** The simplest example of this is provided by the one–loop four–point 1PI graph in the $\varphi_4^4$ theory. The momentum–space integral reads ($p$ is the total external momentum incoming at one vertex and going out of the second vertex of the graph)

$$\begin{aligned} J(p) &= \frac{\hbar^{-1}}{(2\pi)^4} \int d^4k\, \tilde{\Delta}(p+k)\tilde{\Delta}(-k) \\ &= \frac{\hbar}{(2\pi)^4} \int d^4k \frac{1}{[(p+k)^2+m^2][k^2+m^2]} \ . \end{aligned} \tag{2.46}$$

It is logarithmically divergent (i.e., $d(\gamma) = 0$): this singularity is related to the fact that (2.46) is the Fourier transform of $(\Delta(x-y))^2$, i.e., of the ill–defined square of the configuration–space propagator.

To give a proper meaning to such expressions is the first task of renormalization theory. There are a few renormalization procedures available on the market. Let us mention for example the BPHZ momentum–space subtraction procedure [24, 25], the configuration space method of Epstein and Glaser [6], the pole subtractions of the dimensionally regularized integrals [35, 36], the analytic regularization approach [37], the renormalization group approach [38, 39, 40].

We shall content ourselves with a rough sketch of the BPHZ momentum–space subtraction method. The idea of this method is to redefine in a consistent way the integrand $I_\gamma(p, k)$ of a divergent integral $J_\gamma(p)$ (cf. (2.39)) by subtracting from it the few first terms of its Taylor expansion in the external momenta $p$, after these subtractions have been performed for the integrands corresponding to the subgraphs of $\gamma$. The aim of this recursive procedure is to lower the degrees of divergence of the integral and of all its subintegrals until they all become negative. This will assure the absolute convergence of the resulting integral according to the Power–Counting Theorem 2.3.

**Example 2.5** Let us illustrate this with the one–loop example of the integral (2.46). Although this integral is logarithmically divergent, the expression

$$\frac{\partial}{\partial p^\mu} J(p) = -\frac{\hbar}{(2\pi)^4} \int d^4k \, \frac{2(p+k)_\mu}{[(p+k)^2 + m^2]^2 [k^2 + m^2]} \,, \tag{2.47}$$

obtained by a formal differentiation, is convergent. We shall take the latter as the *definition* of the renormalized integral $J(p)$. It is clear that this definition is ambiguous since adding to any particular solution of the problem an arbitrary constant obviously yields another solution.

A particular solution $\hat{J}(p)$ is readily obtained by subtracting from the integrand of (2.46) its value at $p = 0$:

$$\hat{J}(p) = \frac{\hbar}{(2\pi)^4} \int d^4k \left\{ \frac{1}{[(p+k)^2 + m^2][k^2 + m^2]} - \frac{1}{[k^2 + m^2]^2} \right\} \,. \tag{2.48}$$

The latter integral indeed has a negative degree of divergence (equal to $-1$), hence it is convergent, and its derivative coincides with (2.47). An explicit calculation yields (cf. p.229 of Ref. [41])

$$\hat{J}(p) = -\frac{\hbar}{8\pi^2} \left( \frac{1}{2} \sqrt{\frac{4m^2 + p^2}{p^2}} \log \frac{\sqrt{4m^2 + p^2} + \sqrt{p^2}}{\sqrt{4m^2 + p^2} - \sqrt{p^2}} - 1 \right) \,. \tag{2.49}$$

The general solution is given by

$$J(p) = \hat{J}(p) + c \,, \tag{2.50}$$

where $c$ is an arbitrary integration constant. The constant $c$ can be considered as the effect of an arbitrary (finite) counterterm

$$\int d^4x \, c\varphi^4 \tag{2.51}$$

in the action. This counterterm appearing in the one–loop computation is of order $\hbar$.

*Remark.* The subtraction term in the integrand of the right–hand side of (2.48) taken alone would lead to a divergent integral. To understand in which sense "this divergent quantity cancels the divergence of the integral (2.46)", one usually introduces a regularization [42, 43]. The effect of the latter is to render the integral convergent, but depending on a cut–off $\Lambda$. The renormalization is performed through the addition of a counterterm of the form (2.51) to the action, with a coefficient now depending on the cut–off. This cut–off dependence is chosen in such a way that the sum of the regularized integral and of the counterterm contribution is finite in the limit of infinite cut–off. This is the method of "infinite counterterms", which has been shown [44] (in the case of massive theories, within a Pauli–Villars regularization) to be equivalent to the method of momentum space subtractions we are sketching in the present section. Within the regularization method the arbitrary finite coefficient $c$ of the counterterm (2.51) must be viewed as the consequence of the freedom one has to add any cut–off independent number to the cut–off dependent coefficient of the counterterm.

More generally, but still at the order $\hbar$ (i.e., one–loop), let us consider a graph $\gamma$ leading to a divergent integral (2.39), with a degree of divergence $d(\gamma) \geq 0$. The renormalized integral will be defined by its momentum derivatives of order $d(\gamma) + 1$ which correspond to convergent integrals.

A particular convergent integral solution is obtained by subtracting from the integrand $I_\gamma$ its Taylor expansion in the external momenta up to the order $d_{\max}(\gamma)$:

$$\hat{J}_\gamma(p) = \int d^D k R_\gamma(p, k) \, ,$$
$$R_\gamma(p, k) = (1 - t_p^{\delta(\gamma)}) I_\gamma(p, k) \, , \quad \delta(\gamma) = d_{\max}(\gamma) \, , \tag{2.52}$$

where

$$t_p^\omega f(p_1, \cdots, p_n) = \sum_{l=0}^\omega \frac{1}{l!} p_{i_1}^{\mu_1} \cdots p_{i_l}^{\mu_l} \left[ \frac{\partial}{\partial p_{i_1}^{\mu_1}} \cdots \frac{\partial}{\partial p_{i_l}^{\mu_l}} f(p_1, \cdots, p_n) \right]_{p_i = 0} . \tag{2.53}$$

*Remark.* It would have been sufficient to subtract the Taylor expansion up to the order given by the actual degree of divergence $d(\gamma)$. We nevertheless choose to subtract up to the order given by the bound $d_{\max}(\gamma)$ (see (2.42)) In this way the number of subtractions depends only on the external legs of graph, and not on its order.

The general solution of the problem, i.e., a convergent expression whose derivatives of order $d(\gamma)$ coincide with those of the particular solution above, is given by

$$J_\gamma(p) = \hat{J}_\gamma(p) + \sum_{l=0}^{\delta(\gamma)} \frac{1}{l!} c_{\mu_1 \cdots \mu_l} p_{i_1}^{\mu_1} \cdots p_{i_l}^{\mu_l} \, . \tag{2.54}$$

The arbitrary coefficients $c_{\mu_1 \cdots \mu_l}$ represent counterterms of order $\hbar$ in the action, which are local field polynomials with $l$ derivatives. In a power–counting renormalizable theory it is easy to see that the dimensions of the counterterms generated this way do not exceed $D$, the space–time dimension. This means that, if the starting classical Lagrangian is the most general one of dimension bounded by $D$, these coefficients represent (finite) renormalizations of the parameters of the classical theory. Their values have to be fixed by normalization conditions and also, in the presence of symmetries, by invariance requirements. The following example illustrates the preceding considerations for the $\varphi_4^4$ theory.

**Example 2.6** In the $\varphi_4^4$ theory, beyond the logarithmically divergent four–point vertex graphs, the two–point vertex graphs are quadratically divergent (i.e., $d(\gamma) = 2$).

The one–loop approximation was calculated in Example 2.1 (see (2.13)). The subtracted graph is obviously vanishing since the integrand is $p$–independent – non–trivial integrals appear in higher order. Nevertheless there are in principle ambiguities or counterterms, here a second order polynomial $a + bp^2$, with arbitrary constants $a$ and $b$, corresponding to the mass and kinetic counterterms

$$\frac{1}{2} \int d^4x \left( a\varphi^2 + b\partial^\mu \varphi \partial_\mu \varphi \right) . \tag{2.55}$$

The coefficients $a$, $b$ and as well the coefficient $c$ of the quartic counterterm (2.51) may be fixed by the following normalization conditions on the 2–point and 4–point vertex functions in momentum–space[4]:

$$\begin{aligned}
\tilde{\Gamma}_{\varphi\varphi}(p^2)\Big|_{p^2=-m^2} &= 0 , \\
\frac{d}{dp^2} \tilde{\Gamma}_{\varphi\varphi}(p^2)\Big|_{p^2=\mu^2} &= 1 , \\
\tilde{\Gamma}_{\varphi\varphi\varphi\varphi}(p_1 \cdots p_4)\Big|_{p_i^2=\mu^2,\ (p_i+p_j)^2=\frac{4}{3}\mu^2\ (i\neq j)} &= g ,
\end{aligned} \tag{2.56}$$

where $\mu$ is some "normalization mass". These conditions – in the one–loop approximation – fix $a$ and $b$ to zero and $c$, up to a numerical factor, to the value of the subtracted integral (2.48) at the normalization point defined in the third of the equations (2.56).


*Remarks.*

1. The choice of the normalization conditions is free, provided it does lead to a system of nonsingular equations which fix uniquely all the parameters. Since perturbation theory is a formal power series in $\hbar$, it is necessary and sufficient that the normalization conditions, when applied to the tree approximation, admit a unique solution. This tree approximation solution will then reproduce the parametrization of the classical action we have started with.

2. The two–point vertex function being the inverse of the propagator in momentum space – see (2.20) for the configuration space version of this statement – the first normalization condition defines the mass square $m^2$ as the pole of the renormalized propagator, i.e., $m$ as the *physical mass*. Alternative normalization conditions are possible: then the mass parameter $m$ is not the physical mass, but a (calculable) function of the latter and of the coupling constant.

In the case of a theory with composite operators $Q^p$ coupled with external fields $\rho_p$ (see Sect. 2.1), the presence of counterterms linear in $\rho_p$ means a renormalization of $Q^p$. Thus a quantum operator is defined modulo counterterms:

$$Q^p(x) = \hat{Q}^p(x) + \sum_A r_A R^{pA}(x) . \tag{2.57}$$

---

[4]The Fourier transform of a translation invariant function is defined in (2.34).

Here $\hat{Q}^p$ is the particular[5] solution defined by the corresponding BPHZ–subtracted integral $\hat{J}$ (2.52). The quantities $R^{pA}$ are all local field polynomials of dimensions not exceeding the dimension of $Q^p$ if the theory is power–counting renormalizable. They are furthermore selected by the symmetries which are preserved by the subtraction scheme (for example Poincaré invariance). The arbitrary constants $r_A$ are fixed by normalization conditions on the inserted vertex functional $Q^p(x) \cdot \Gamma$, and by other possible symmetry requirements.

### 2.2.3  Generalization to Higher Loop Graphs

If a multiloop divergent graph $\gamma$ has all its proper 1PI subgraphs $\tau$ convergent, i.e., if

$$d(\gamma) \geq 0 \text{ and } d(\tau) < 0 \quad \forall \tau \subset \gamma \, , \, \tau \neq \gamma$$

then the formulas above apply.

If subgraphs are divergent, then they must be subtracted like in (2.52) before the overall graph $\gamma$ is subtracted. These subtractions are thus recursively performed, beginning with the innermost subgraphs and then continuing up to the overall graph. This offers no difficulty if there are no overlapping divergent subgraphs. If there are overlapping divergent subgraphs, Zimmermann's forest formula [24, 45, 46, 47] must be applied. This is the BPHZ subtraction scheme, which generalizes the procedure we have used in our one loop example. It consists in replacing the integrand $I_\gamma$ of the integral (2.39) by a subtracted integrand $R_\gamma$ whose integral

$$J_\gamma^R(p) = \left[ \int \prod_{l=1}^L d^D k_l \right] R_\gamma(p, k) \tag{2.58}$$

is well–defined.

Before stating the general results, let us briefly describe the BPHZ scheme, in the case of massive fields. A *forest U* of a 1PI graph $\gamma$ is a set of divergent 1PI subgraphs $\lambda$ of $\gamma$ such that any two elements of $U$ are either disjoint or one is a subgraph of the other. The set of all forests of $\gamma$ is denoted by $\mathcal{F}$. It includes the empty forest $\emptyset$ and the trivial forest $\{\gamma\}$ consisting of the graph $\gamma$ alone.

The forest formula yielding the renormalized integrand in (2.58) reads, for a 1PI graph $\gamma$,

$$R_\gamma(p, k) = \sum_{U \in \mathcal{F}(\gamma)} \prod_{\lambda \in U} \left( -t_{p_\lambda}^{\delta(\lambda)} \right) I_\gamma(p, k) \, . \tag{2.59}$$

$t_{p_\lambda}^{\delta(\lambda)}$ is the Taylor operator (2.53) to the order $\delta(\lambda)$ with respect to the external momenta $p_\lambda$ of the subgraph $\lambda$. By convention this operator is equal to 0 if $\delta(\lambda) < 0$. The summation runs over the set $\mathcal{F}$ of all forests $U$ of $\gamma$, the product runs over all the elements of $U$, ordered in such a way that $t_{p_{\lambda_2}}^{\delta(\lambda_2)}$ stays at the right of $t_{p_{\lambda_1}}^{\delta(\lambda_1)}$ if $\lambda_2$ is a subgraph of $\lambda_1$.

If the divergent subgraphs are nonoperating they form a forest $U^*$, and the forest formula takes the simpler form

$$R_\gamma(p, k) = \prod_{\lambda \in U^*} \left( 1 - t_{p_\lambda}^{\delta(\lambda)} \right) I_\gamma(p, k) \, . \tag{2.60}$$

---

[5]This definition of $\hat{Q}^p$ is the normal product $N[Q]$ of Zimmermann [45].

The meaning of the latter formula is rather obvious: each divergent subgraphs is subtracted with respect to its own external momenta, as in (2.52) for a one–loop graph, in an iterative procedure which begins with the most inner divergent subgraph.

Coming back to the general case with possible overlapping divergent subgraphs, we may mention that Zimmermann's forest formula (2.59) is the explicit solution of a recurrence procedure due to Bogoliubov [5, 42], called the *operation R*. Let us sketch the latter briefly. Denoting by $\gamma$ the 1PI diagram to be defined, one first writes

$$R_\gamma = (1 - T_\gamma)\bar{R}_\gamma .   \tag{2.61}$$

$R_\gamma$ and $\bar{R}_\gamma$ represent respectively the renormalized integrand and the one where the contributions of all the divergent proper subgraphs have already been subtracted. $T_\gamma$ is a subtraction operator on $\bar{R}_\gamma$, which may be the momentum space Taylor operator (2.53), or the pole subtraction operator of the dimensional renormalization scheme [36, 43], etc. By convention $T_\lambda = 0$ if the graph or subgraph $\lambda$ has a strictly negative superficial degree of divergence.

$\bar{R}_\gamma$ itself is recursively defined by the formula

$$\bar{R}_\lambda = I_\lambda + \sum_{\substack{\{\lambda_1,\cdots,\lambda_s\} \\ \lambda_i \cap \lambda_j = \emptyset}} I_{\lambda/\{\lambda_1,\cdots,\lambda_s\}} \prod_{i=1}^{s} \left( -T_{\lambda_i} \bar{R}_{\lambda_i} \right) .   \tag{2.62}$$

$\lambda$ is any 1PI subgraph of $\gamma$, or $\gamma$ itself, $I_\lambda$ is the unsubtracted integrand of $\lambda$. The sum extends to all sets of mutually disjoint proper 1PI subgraphs $\lambda_i$ of $\lambda$. $I_{\lambda/\{\lambda_1,\cdots,\lambda_s\}}$ represents the unsubtracted integrand corresponding to the reduced graph $\lambda/\{\lambda_1,\cdots,\lambda_s\}$ obtained by shrinking to a point each of the subgraphs $\lambda_i$.

### 2.2.4 General Results

The general results can be summarized in a set of three statements:

(a) Write the contributions to the Green functions at a given order as momentum space integrals according to the Feynman rules.

(b) Each UV divergent momentum space integral is given a meaning by subtracting its integrand according to the BPHZ forest formula (2.59), which generalizes (2.48) and (2.52). The ambiguity of the subtraction procedure will result in a set of finite counterterms, such as (2.51) and (2.55), each one corresponding to a given type[6] of divergent 1PI diagrams.

(c) The coefficients of the counterterms are fixed by the symmetries, if any, of the theory – this will be explained at length in the rest of the book – and by a set of normalization conditions, such as (2.56), on the vertex functions to which divergent diagrams contribute.

These statements define what is called a *renormalization procedure*, which in fact is equivalent to a proper definition of the theory.

---

[6]The type of a diagram is defined by the nature and the number of its external legs.

*Remarks.*

1. The renormalization procedure is performed order by order in $\hbar$. The coefficient of a counterterm corresponding to a diagram of order $n$ is of order $n$.

2. Subtractions performed – in the one–loop case according to the formula (2.52), and according to the forest formula (2.59) in the general case – i.e., up to the order given by the bound $d_{\max}$ of the actual degree of divergence, are called *minimal*. Oversubtractions are possible. They consist in replacing the dimension $d_Q$ of $Q$ in Eq. (2.43) by a number $\delta_Q > d_Q$. The insertion constructed in this way is called [45] a *nonminimal normal product* of degree $\delta_Q$.

3. If propagators of massless fields are present, the subtraction terms of certain integrals may lead to infrared (IR) divergences, i.e., to singularities caused by an inappropriate behaviour of the integrands at zero momentum. The subtraction rules must then be modified [25]. Without entering into the details, we only mention that there exists an extension of the BPHZ momentum space subtraction scheme which takes into account the possible infrared singularities caused by the massless propagators. This scheme, known as the BPHZL subtraction procedure [25, 47], ensures the existence of the Green functions as tempered distributions in the general case.

4. The use of the BPHZ subtraction scheme is of course not mandatory. Any well defined procedure is acceptable and leads to the same results [48], provided one uses a complete set of normalization conditions, making the Green functions uniquely defined.

5. We are concerned only with Green functions, i.e., with vacuum expectation values of time ordered products of fields. The generalization to Wightman functions and, more generally, to vacuum expectation values of products of time–ordered and anti–time–ordered products, does exist. It can be found in [49].

### 2.2.5 Renormalizability and Nonrenormalizability

We have seen, in the case of the $\varphi_4^4$ theory in 4–dimensional space–time, that all the ambiguities resulting from the renormalization procedure correspond to the renormalization of the coefficients of the terms of the classical action. In such a situation the theory is called *renormalizable*[7]. We remark that the dimension of the counterterms is bounded by the space–time dimension.

In the general case there are three possibilities, according to the behaviour of the superficial divergence degree (2.41):

$$d(\gamma) = D + \sum_{\substack{\text{all vertices } V_k}} (\dim(V_k) - D) - \sum_i N_i d_i \ ,$$

with respect to the order of perturbation theory:

---

[7]One often encounters the expression *multiplicatively renormalizable*.

1. The number of divergent graph types (the type of a graph is characterized by the nature and the number of its external legs) increases indefinitely with the order. This may have either one of the following two causes:

   i) The presence of interaction vertices $V_k$ of dimension greater than the space–time dimension $D$. The resulting ambiguities then correspond to counterterms of ever increasing dimension, hence to an indefinite number of free arbitrary parameters. The theory is then said to be *power–counting non-renormalizable*. Since the number of interactions, i.e., of coupling constants, is indefinite, every predictability is loosed, the theory has no physical meaning and is simply said to be *nonrenormalizable*.

   ii) The presence of fields $\Phi_i$ of dimension $d_i = 0$. If there is no interaction vertex of dimension greater than $D$, the dimension of the counterterms remains bounded by $D$, although their number is indefinite since they may contain arbitrary powers of the dimensionless fields. Such a theory belongs to the class of the *power–counting renormalizable* theories, although it may be nonrenormalizable, i.e., unphysical, since it may depend on an indefinite number of physical parameters (see Subsect. 2.2.6).

2. The number of divergent graph types is bounded. According to the expression above for the degree of divergence, this forbids the presence of any interaction vertex of dimension greater than $D$. The number of counterterms is thus finite and their dimension is bounded by $D$. The theory is also *power–counting renormalizable*.

3. The number of types of divergent diagrams is not only bounded, but there are divergent diagrams only up to a certain maximal order. Thus their total number itself is finite. The theory is called *superrenormalizable*. This happens if all interaction vertices have a dimension lower than the space–time dimension $D$.

### 2.2.6 Physical and Nonphysical Renormalizations

In any of the situations depicted above one has to distinguish between *physical renormalizations*, which affect physical parameters such as the coupling constants and the particle masses, and *nonphysical renormalizations*, which correspond to a mere redefinition of the field variables[8], as we shall see in the foregoing example.

**Example 2.7** The three independent counterterms of the $\varphi_4^4$ theory may be chosen as

$$a'\partial_{m^2}S + b'NS + c'\partial_g S = \int d^4x \left( b'\partial^\mu\varphi\partial_\mu\varphi + (\frac{1}{2}a' + m^2 b')\varphi^2 + (\frac{g}{3!}b' + \frac{1}{4!}c')\varphi^4 \right),$$
$$(2.63)$$

where $S$ is the action (2.1) and $N$ is the "counting operator"

---

[8] In gauge theories (See Chap.4) the free parameters correspond to BRS invariant counterterms. The BRS invariant counterterms corresponding to nonphysical renormalizations are characterized by being BRS variations.

$$N = \int d^4x \; \varphi(x) \frac{\delta}{\delta\varphi(x)} \; . \tag{2.64}$$

One sees that the parameter $b'$ can be eliminated by the field redefinition[9] ("wave function renormalization")

$$\varphi \quad \rightarrow \quad (1 - b')\varphi + O(b'^2). \tag{2.65}$$

On the other hand $a'$ and $c'$, which affect the mass and the coupling constant, are physical renormalizations.

It is evident that a theory depending on an indefinite number of parameters will still be a good theory, i.e., it will possess a predictive power, if the number of physical parameters itself is finite.

### 2.2.7 The Role of Symmetry

The discussion up to now has relied only on considerations of power–counting. It has been based on the formula (2.41) for the superficial degree of divergence, which gives all the possibly infinite graphs, i.e., all the possible counterterms. However restrictions imposed to the theory may limit the number of these counterterms. Such restrictions most often come from *symmetries*[10].

**Example 2.8** This is the case for the $\varphi_4^4$ theory: the absence of a counterterm $\varphi^3$ is due to the invariance of the theory under the $G$–parity transformation

$$\varphi \quad \rightarrow \quad -\varphi \; , \tag{2.66}$$

which forbids diagrams with an odd number of external legs.

For a theory to be physically meaningful, it is enough that the number of physical parameters (as defined in Subsect. 2.2.6) be finite once all symmetry requirements are fulfilled. Such a theory will be called a *renormalizable theory*[11].

On the other hand, power–counting renormalizable theories with an indefinite number of counterterms, may depend only on a finite number of physical ones, due to the symmetries. Important examples are provided by the supersymmetric gauge theories [26] and by the nonlinear sigma models [52, 53, 54].

---

[9]The coefficients of the counterterms have to be considered as infinitesimal since they concern a certain order $n$ of a formal power series.

[10]There is a more general way to impose relations between coupling constants known as "reduction of couplings" [50, 51]. There are examples of solutions of the reduction problem which do not seem to correspond to any symmetry [51]

[11]There is no known example of a power–counting nonrenormalizable theory – as defined by the point 1.i) in Subsect. 2.2.5 – in which symmetry requirements would reduce the number of physical counterterms to a finite amount.

## 2.3 Asymptotic Behaviour

The subtracted integral (2.48), (2.49) in the example discussed in Subsect. 2.2.2 behaves as $\log p$ in the region of infinite momentum:

$$\hat{J}(p) = -\frac{\hbar}{16\pi^2}\left(\log\frac{p^2}{m^2} - 2\right) + O\left(\frac{1}{p^2}\log p^2\right) . \tag{2.67}$$

The asymptotic behaviour of generic Feynman graph integrals is given by Theor. 2.10 below. Before stating the theorem we need the following definition:

**Definition 2.9** *The set of external momenta $\{p_1, p_2, \cdots\}$ entering a diagram is said to be nonexceptional [55] if no (nontrivial) partial sum of them vanishes.*

**Theorem 2.10** *(Weinberg Theorem [33]) Let us assume the theory to be power-counting renormalizable, and let $J_\gamma^R(p)$ be the momentum space contribution of a 1PI graph $\gamma$, with divergence degree $d(\gamma)$, to the vertex function (2.35), subtracted if necessary, cf. (2.58), but minimally subtracted[12]. Provided the momenta $p_1, p_2, \cdots$ entering the diagram are nonexceptional, the UV asymptotic behaviour of this contribution has the form*

$$J_\gamma^R(\lambda p_1, \cdots, \lambda p_{N_1}, \lambda p_{N_1+1}, \cdots, \lambda p_{N_1+N_2}, \cdots) \sim (\log\lambda)^\alpha \lambda^{D-d(\gamma)} , \qquad \lambda \to \infty , \tag{2.68}$$

*where $\alpha$ is a number depending on the number of loops of the diagram $\gamma$.*

Since the divergence degree of a diagram contributing to the vertex function (2.35) is bounded by $d_{\max}(\gamma)$ according to (2.42), one can state:

**Corollary 2.11** *The high momentum behaviour of the vertex function (2.35) in a power–counting renormalizable theory is given by*

$$\tilde{\Gamma}_{N_1 N_2 \cdots}(\lambda p_1, \cdots, \lambda p_{N_1}, \lambda p_{N_1+1}, \cdots, \lambda p_{N_1+N_2}, \cdots) \sim (\log\lambda)^\alpha \lambda^{D-d_{\max}(\gamma)} , \qquad \lambda \to \infty , \tag{2.69}$$

*where the number $\alpha$ depends on the order.*

---

[12]Minimal subtractions are defined in the third remark at the end of Subsect. 2.2.2.

# 3. Quantum Action Principle and Ward Identities

The symmetries of a quantum field model are expressed by Ward identities, i.e., by certain identities which the Green functions of the theory must obey. The aim of the renormalization procedure is to prove the renormalizability of a given symmetry of the corresponding classical theory. This means that one has to construct a theory of Green functions fulfilling perturbatively, to all orders, the Ward identities which express this symmetry. We shall explain in this chapter how to write the Ward identities in the functional formalism, beginning with the classical theory – equivalent to the tree approximation of the quantum theory – and then we shall proceed to the extension to all orders of perturbation theory, for very simple models with rigid symmetry. The main tool being the renormalized [12, 13, 14] version of the Schwinger action principle [56], the *quantum action principle* (QAP), a section will be devoted to the latter.

In view of the more sophisticated models – in particular gauge theories – which will be studied later, we shall introduce the notion of BRS symmetry already in the case of rigid invariance.

As we have already mentioned in the Introduction, a given ultraviolet subtraction scheme may break the symmetries of the problem. This will occur in particular if there is no invariant regularization procedure available. It turns out however that the properties of locality and power–counting of the possible breakings are fully characterized by the QAP. The simple examples treated in this chapter will explicitly show how these breakings may then be eliminated by the introduction of compensating noninvariant local counterterms.

We shall also meet a symmetry which cannot be restored in this way, namely scale invariance. This will be our first encounter with an "anomaly". The breaking of scale invariance by the scale anomaly will be expressed by a parametric partial differential equation known as the Callan–Symanzik equation [57].

## 3.1 Ward Identities in the Tree Approximation

Let us consider a theory given by the Lagrangian (2.4) of Sect. 2.1 involving the fields $\phi_i$, and let us suppose that the the corresponding action (2.3), $S(\phi)$, is invariant under the infinitesimal transformations

$$\delta\phi_i(x) = i\varepsilon^a R_{ai}(x) \ . \tag{3.1}$$

The functions $R_{ai}(x)$, $a = 1, 2, \cdots$ are local functionals, i.e., polynomials or more generally formal power series, of the fields and of their derivatives. The infinitesimal parameters $\varepsilon^a$ are constant: in this chapter we restrict the discussion to *rigid* transformations. We assume that the transformations (3.1) belong to a representation of a Lie group $G$. More precisely, calling $X_a$ the elements of a basis of the Lie algebra Lie $G$:

$$[X_a, X_b] = if_{ab}{}^c X_c ,  \tag{3.2}$$

where the $f_{ab}{}^c$'s are the structure constants, obeying the Jacobi identity

$$\sum_{\text{cyclic perm. of } abc} f_{ab}{}^d f_{dc}{}^e = 0 ,  \tag{3.3}$$

we demand that the $R_{ai}(x)$ obey the equations

$$\int dy \left( R_{bj}(y) \frac{\delta R_{ai}(x)}{\delta \phi_j(y)} - R_{aj}(y) \frac{\delta R_{bi}(x)}{\delta \phi_j(y)} \right) = if_{ab}{}^c R_{ci}(x) .  \tag{3.4}$$

The infinitesimal variation of a field functional $F(\phi)$ reads

$$\delta F = -i\varepsilon^a \mathcal{W}_a F ,$$
$$\mathcal{W}_a := -\int dx\, R_{ai} \frac{\delta}{\delta \phi_i} ,  \tag{3.5}$$

and, due to (3.4), the functional operators $\mathcal{W}_a$ obey the Lie algebra commutation relations (3.2):

$$[\mathcal{W}_a, \mathcal{W}_b] = if_{ab}{}^c \mathcal{W}_c .  \tag{3.6}$$

*Remark.* The $-$ sign in the definition of the operators $\mathcal{W}_a$ has been chosen in such a way that they obey the same Lie algebra as the generators $X_a$ of the group.

### 3.1.1 Linear Transformations

In the examples we are going to study in this chapter the transformations (3.1) are linearly homogeneous in the fields. In this case the functions $R_{ai}$ take the form

$$R_{ai}(x) = T_{ai}{}^j \phi_j(x) ,  \tag{3.7}$$

with the matrices $T_a$ obeying the Lie algebra commutation rules (3.2). The invariance of the classical action $S(\phi)$ is expressed, in the linear case, by the identities

$$\mathcal{W}_a S := -\int dx\, T_{ai}{}^j \phi_j \frac{\delta}{\delta \phi_i} S = 0 .  \tag{3.8}$$

Let us now remember that the classical action coincides with the generating functional of the vertex functions in the tree approximation. Its Legendre transform, defined in Sect. 2.1 by (2.17) and (2.18), is the generating functional $Z^{c(0)}(J)$ of the connected Green functions in the tree approximation, and its exponential $Z^{(0)} = \exp -Z^{c(0)}/\hbar$ generates the Green functions, in the same approximation. It is a simple exercise to show that the invariance condition (3.8) is equivalent, for the tree approximation Green functional, to the identities

$$W_a Z^{(0)}(J) = \int dx J^i T_{ai}{}^j \frac{\delta}{\delta J^j} Z^{(0)}(J) = 0 \ . \tag{3.9}$$

For the individual Green functions (see (2.5), (2.6)) in the tree approximation, this yields

$$\sum_{n=1}^{N} \left\langle T\phi_{i_1}(x_1) \cdots \phi_{i_{n-1}}(x_{n-1}) T_{ai_n}{}^{j_n} \phi_{j_n}(x_n) \phi_{i_{n+1}}(x_{n+1}) \cdots \phi_{i_N}(x_N) \right\rangle^{(0)} = 0 \ . \tag{3.10}$$

Such relations between Green functions, due to a symmetry, are called *Ward identities*.

### 3.1.2 Nonlinear Rigid Invariance: BRS Transformations

In the more general case of functions $R_{ai}(x)$ which are nonlinear in the fields[1] – and which thus are subjected to be renormalized – one is led to couple these functions to external fields $\rho^i(x)$ through the addition of a term

$$S_{\text{ext}}(\phi, \rho) = i \int dx \rho^i \varepsilon^a R_{ai} \tag{3.11}$$

to the original action $S(\phi)$. Let us write the total classical action as

$$\Gamma^{(0)} = S(\phi) + S_{\text{ext}}(\phi, \rho) \ , \tag{3.12}$$

where the notation reminds us that the classical action coincides with the vertex functional in the zeroth order, i.e., tree graph, approximation. We would like to derive Ward identities as in the linear case. Unfortunately this action is not invariant under the transformations (3.1) because of the external field piece. The way out [58, 52] is to

i) take the infinitesimal parameters $\varepsilon^a$ as Grassmann numbers, i.e., anticommuting numbers: $\{\varepsilon^a, \varepsilon^b\} = 0$,

ii) let these parameters transform, too:

$$\delta\varepsilon^a = -\frac{1}{2} f_{bc}{}^a \varepsilon^b \varepsilon^c \ , \tag{3.13}$$

where the $f_{bc}{}^a$ are the Lie group structure constants defined in (3.2), (3.4).

After this redefinition of the transformation rules, the operator $\delta$ defined by (3.1) and (3.13) is nilpotent:

$$\delta^2 = 0 \ , \tag{3.14}$$

and the total action is now invariant: $\delta\Gamma^{(0)} = 0$. The latter invariance can be written in a functional form as

$$\mathcal{S}(\Gamma^{(0)}) = \int dx \frac{\delta\Gamma^{(0)}}{\delta\rho^i} \frac{\delta\Gamma^{(0)}}{\delta\phi_i} - \frac{1}{2} f_{bc}{}^a \varepsilon^b \varepsilon^c \frac{\partial\Gamma^{(0)}}{\partial\varepsilon^a} = 0 \ . \tag{3.15}$$

---

[1] The typical example is that of the BRS symmetry of the nonabelian gauge theories. It will be studied in Chap. 4.

The nilpotent transformation $\delta$ introduced here is called a *BRS operator*, and the last equation a *Slavnov–Taylor identity*. Although the BRS formalism was first introduced [15, 16, 59] in order to deal with the quantization of gauge theories, i.e., of theories with local invariances, it may be very useful also for theories with rigid symmetries [58]. The geometrical interpretation of this BRS operator as the exterior derivative on the group manifold is reviewed in [60] for rigid symmetries as well as for local symmetries.

Let us conclude this Section by the following very important observation [16]. Without referring to any *a priori* given group, we may consider (3.1) and (3.13) as Ansätze, with arbitrary functions $R_{ai}$ of the fields and of their derivatives, and with arbitrary coefficients $f_{bc}{}^a$, taking the parameters $\varepsilon^a$ as Grassmann numbers. We may then require the operation $\delta$ to be nilpotent. From $\delta^2\varepsilon = 0$, we shall obtain the Jacobi identity (3.3), and from $\delta^2\phi = 0$ the condition (3.4). The Jacobi identity allows us to identify the coefficients $f_{bc}{}^a$ as the structure constants of a certain Lie group. Then the identities (3.4) show that the variations (3.1) are infinitesimal transformations of the fields in a certain representation of this group.

## 3.2 The Quantum Action Principle

The free Green functional (2.8) obeys the equation

$$\hbar K^{ij}\frac{\delta Z_{\text{free}}}{\delta J^j(x)} - J^i(x)Z_{\text{free}} = 0 \;, \tag{3.16}$$

which is nothing else than the free field equation written in functional form.

A *formal* field equation for the interacting theory can be derived from the formal expression (2.10) of the Green functional. Indeed a simple calculation

$$\begin{aligned}
\hbar K^{ij}\frac{\delta Z}{\delta J^j(x)} &= \mathcal{N}e^{-\frac{1}{\hbar}S_{\text{int}}\left(-\hbar\frac{\delta}{\delta J}\right)}J^i(x)Z_{\text{free}} \\
&= J^i(x)Z + \left[\mathcal{N}e^{-\frac{1}{\hbar}S_{\text{int}}\left(-\hbar\frac{\delta}{\delta J}\right)}, J^i(x)\right]Z_{\text{free}} \\
&= J^i(x)Z + \left.\frac{\delta S_{\text{int}}}{\delta\phi_i(x)}\right|_{\phi\to-\hbar\frac{\delta}{\delta J}}Z
\end{aligned} \tag{3.17}$$

leads to the equation

$$\left[K^{ij}\phi_j(x) + \frac{\delta S_{\text{int}}}{\delta\phi_i(x)}\right]_{\phi\to-\hbar\frac{\delta}{\delta J}}Z + J^i(x)Z = 0 \;, \tag{3.18}$$

or:

$$\left.\frac{\delta S}{\delta\phi_i(x)}\right|_{\phi\to-\hbar\frac{\delta}{\delta J}}Z + J^i(x)Z = 0 \;, \tag{3.19}$$

where $S$ is the total action (2.3). The first term of the last equation is a Green functional with the insertion of the composite field $\delta S/\delta\phi_i$, which has dimension $D-d_i$.

Of course the derivation above as well as the result are quite formal since we have not taken into account the problem of the UV divergences. But it has been shown [12] that the structure of (3.19) survives the renormalization procedure, if the theory is power–counting renormalizable – power–counting renormalizability will be assumed for the rest of these lectures. The renormalized version of (3.19) reads

$$J^i(x)Z + \Delta^i(x) \cdot Z = 0 , \qquad (3.20)$$

where $\Delta^i(x)$ is a *local* composite field operator of dimension bounded by $(D - d_i)$, coinciding with the "classical insertion", i.e., the classical field polynomial, $\delta S/\delta \phi_i$ in the classical approximation. In terms of the connected functional this reads

$$J^i(x)Z^c + \Delta^i(x) \cdot Z^c = 0 , \qquad (3.21)$$

and, after a Legendre transformation,

$$\frac{\delta \Gamma}{\delta \phi_i(x)} - \Delta^i(x) \cdot \Gamma = 0 \qquad (3.22)$$

for the vertex functional.

Multiplying the formal equation (3.19) with a polynomial $Q^a(x)$ of dimension $d_{Q^a}$ – coupled with an external field $\rho_a(x)$ of dimension $D - d_{Q^a}$, (see Sect. 2.1) – one obtains, still formally:

$$\left( Q^a(x) \frac{\delta S}{\delta \phi_i(x)} \right) \cdot Z - \hbar J^i(x) \frac{\delta Z}{\delta \rho_a(x)} = 0 . \qquad (3.23)$$

There is also a rigorous version of this equation for the renormalized theory [12, 13]:

$$- \hbar J^i(x) \frac{\delta Z}{\delta \rho_a(x)} + \Delta^{ai}(x) \cdot Z = 0 , \qquad (3.24)$$

where $\Delta^{ai}(x)$ is a local composite field operator of dimension bounded by $(D-d_i+d_{Q^a})$. For the vertex functional this becomes

$$\frac{\delta \Gamma}{\delta \rho_a(x)} \frac{\delta \Gamma}{\delta \phi_i(x)} = \Delta^{ai}(x) \cdot \Gamma . \qquad (3.25)$$

Equations (3.24) and (3.25) are the main content of the quantum action principle [12, 13, 14] (ref. [14] concerns the generalization of the BPHZ renormalization of composite field operators and of the quantum action principle to theories where some fields are massless). The general form (3.24) or (3.25) of the quantum action principle will be needed in the study of symmetries with *nonlinear* transformation laws as in (3.5).

Note that, due to (2.23) and (2.33), (3.25) is in fact the quantum extension of the tautological classical identity

$$\frac{\delta S}{\delta \rho_a(x)} \frac{\delta S}{\delta \phi_i(x)} = Q^a(x) \frac{\delta S}{\delta \phi_i(x)} . \qquad (3.26)$$

In case the field $Q^a$ is linear in the basic fields $\phi_i$, no external field $\rho^a$ needs to be introduced, and (3.25) takes the simpler form

$$Q^a(x) \frac{\delta \Gamma}{\delta \phi_i(x)} = \Delta^{ai}(x) \cdot \Gamma , \qquad (3.27)$$

with the same conditions on the dimensions of the insertion $\Delta^{ai}$.

Another useful result, also part of the quantum action principle, is obtained by varying the Green functions with respect to a parameter $\lambda$ of the theory (mass, coupling constant, renormalization mass $\mu$, etc. ). It writes

$$-\hbar\frac{\partial}{\partial\lambda}Z = \Delta \cdot Z \quad \text{or} \quad \frac{\partial}{\partial\lambda}\Gamma = \Delta \cdot \Gamma , \tag{3.28}$$

where $\Delta$ is a space–time integrated local composite field of dimension bounded by $D$. At lowest order the latter reads

$$\frac{\partial}{\partial\lambda}\Gamma = \frac{\partial}{\partial\lambda}S + O(\hbar) . \tag{3.29}$$

The complete proofs of the quantum action principle are available in the original literature[2]. Sketches of the proofs are presented for example in [46, 47, 12, 61]

## 3.3 Basis of Insertions

It is clear from their definition that the insertions of composite field operators form a linear space, and it is important to be able to construct basis of this space. It is remarkable that, in perturbation theory, any such basis of renormalized insertions is completely characterized by the corresponding classical basis:

**Proposition 3.1** *Let*

$$\left\{ \Delta^p \cdot \Gamma = \Delta^p_{\text{class}} + O(\hbar) \mid p = 1, 2, \cdots ; \dim(\Delta^p) \leq \bar{d} \right\} \tag{3.30}$$

*be a set of insertions such that their classical approximations[3] $\Delta^p_{\text{class}}$ form a basis for the classical insertions of dimension bounded by $\bar{d}$. The $\Delta$'s may be $x$–dependent local insertions or space–time integrals thereof.*

*Then the set (3.30) is a basis for the quantum insertions of dimension bounded by $\bar{d}$.*

*Remark.* Any set $\{\Delta'^p \cdot \Gamma\}$ with $\Delta'^p \cdot \Gamma - \Delta^p \cdot \Gamma = O(\hbar)$ is an equivalent basis.

*Proof.* The proof typically relies on the property of perturbation theory being a formal power series in the expansion parameter, here $\hbar$. We want to show that any insertion $\Delta \cdot \Gamma$ of dimension bounded by $\bar{d}$ can be expanded in the basis (3.30). We can write, first,

$$\Delta \cdot \Gamma = \Delta_{\text{class}} + O(\hbar) = \sum_p c_p \Delta^p_{\text{class}} + O(\hbar)$$
$$= \sum_p c_p \Delta^p \cdot \Gamma + O(\hbar) ,$$

---

[2]See quoted references. C.f. also, for example, [36] for a discussion in the framework of dimensional regularization, [37] in the framework of analytic regularization and [40] in the renormalization group approach.

[3]See (2.33).

the second equality following from the hypothesis. The proof then proceeds by induction. Let us assume the desired result to hold up to the order $n-1$. We can thus write

$$\Delta \cdot \Gamma = \sum_p c_p \Delta^p \cdot \Gamma + \hbar^n \Delta'_{\text{class}} + O(\hbar^{n+1}) ,$$

where the classical insertion $\Delta'_{\text{class}}$ represents the corrections of order $n$. The latter being expanded in the classical basis, we obtain

$$\Delta \cdot \Gamma = \sum_p \left( c_p \Delta^p \cdot \Gamma + \hbar^n c'_p \Delta^p_{\text{class}} \right) + O(\hbar^{n+1})$$
$$= \sum_p (c_p + \hbar^n c'_p) \Delta^p \cdot \Gamma + O(\hbar^{n+1}) ,$$

which is the result at the next order $n$.  ∎

This proposition will be used in particular in Sect. 3.4.2 for the derivation of the Callan–Symanzik equation.

## 3.4 A U(1) Scalar Model

As a first application of the renormalization theory sketched in the preceding sections, and as an introduction to the concept of symmetry in the framework of perturbative quantum field theory, let us study the renormalization of a simple model with a U(1) rigid symmetry.

The model contains a complex scalar field $\varphi$ in four–dimensional space–time and is supposed to be invariant under the phase transformation (written in infinitesimal form)

$$\delta\varphi(x) = i\varepsilon\varphi(x) , \qquad \delta\bar{\varphi}(x) = -i\varepsilon\bar{\varphi}(x) , \qquad (3.31)$$

where $\varepsilon$ is a constant infinitesimal real parameter. The most general invariant, power–counting renormalizable classical action is

$$S = \int d^4x \left( \partial\bar{\varphi}\partial\varphi + m^2\bar{\varphi}\varphi + \frac{g}{4}(\bar{\varphi}\varphi)^2 \right) . \qquad (3.32)$$

This action is moreover invariant under the discrete charge conjugation transformation:

$$C : \quad \varphi \leftrightarrow \bar{\varphi} . \qquad (3.33)$$

### 3.4.1 Ward Identity

U(1)–invariance of the action can be expressed by the classical Ward identity

$$\mathcal{W}S = 0 , \qquad (3.34)$$

where we have introduced the functional differential Ward operator

$$\mathcal{W} = \int d^4x \left( \varphi\frac{\delta}{\delta\varphi} - \bar{\varphi}\frac{\delta}{\delta\bar{\varphi}} \right) . \qquad (3.35)$$

We would like the same invariance condition to hold in the higher orders of perturbation theory:

$$\mathcal{W}\Gamma(\varphi, \bar{\varphi}) = 0 , \tag{3.36}$$

where $\Gamma$ is the vertex functional.

But it very often happens that the subtraction procedure breaks these identities. In such a case the question is: is it possible to adjust some of the counterterms which have been left free after the subtractions, in such a way that the Ward identities are recovered? If the answer is no, then one speaks of an *anomaly*, i.e., of a breaking of the symmetry through the radiative corrections. Let us show that there is no anomaly in the present case.

The proof is recursive. Under the assumption that there exists a vertex functional $\Gamma_{(n-1)}$ obeying the Ward identity (3.36) till the order $n-1$ in $\hbar$,

$$\mathcal{W}\Gamma_{(n-1)} = O(\hbar^n) , \tag{3.37}$$

we shall show that it holds at the next order, provided the free counterterms of order $n$ which are not U(1)–invariant are suitably chosen.

We shall call $S_{(n-1)}$ the action, with all its counterterms till the order $n-1$, which leads to the functional $\Gamma_{(n-1)}$ and to (3.37).

The action principle in the form (3.27) can be applied, with the result:

$$\mathcal{W}\Gamma_{(n-1)} = \hbar^n \Delta \cdot \Gamma = \hbar^n \Delta + O(\hbar^{n+1}) , \tag{3.38}$$

where $\Delta$ is a Poincaré invariant, integrated local insertion of dimension less or equal to four. We have taken (2.33) into account for writing the second equality, where $\Delta$ is a classical local field polynomial. Due to the invariance under the charge conjugation (3.33), and due to the fact that the Ward operator $\mathcal{W}$ (3.35) is odd under charge conjugation, $\Delta$ is odd, as well.

The classical field polynomial $\Delta$, obeying the constraints listed above, can be expanded in the basis:

$$\int d^4x \left(\varphi^2 - \bar{\varphi}^2\right) , \quad \int d^4x \left(\varphi^3 \bar{\varphi} - \bar{\varphi}^3 \varphi\right), \quad \int d^4x \left(\partial\varphi\partial\varphi - \partial\bar{\varphi}\partial\bar{\varphi}\right) ,$$
$$\int d^4x \left(\varphi^4 - \bar{\varphi}^4\right) .$$

These four basis elements can be written as the $\mathcal{W}$–variation of four independent field polynomials, even under charge conjugation and of dimension less or equal to four:

$$\frac{1}{2}\int d^4x \left(\varphi^2 + \bar{\varphi}^2\right) , \quad \frac{1}{2}\int d^4x \left(\varphi^3 \bar{\varphi} + \bar{\varphi}^3 \varphi\right) , \quad \frac{1}{2}\int d^4x \left(\partial\varphi\partial\varphi + \partial\bar{\varphi}\partial\bar{\varphi}\right) ,$$
$$\frac{1}{4}\int d^4x \left(\varphi^4 + \bar{\varphi}^4\right) . \tag{3.39}$$

Denoting the latter field polynomials by $\Delta_i$, we can write (3.38) as

$$\mathcal{W}\Gamma_{(n-1)} = \hbar^n \mathcal{W}\hat{\Delta} + O(\hbar^{n+1}) ,$$
$$\text{where } \hat{\Delta} = \sum_{1}^{4} r^i \Delta_i . \tag{3.40}$$

Let us now replace the action $S_{(n-1)}$ by the new action

$$S_{(n)} = S_{(n-1)} - \hbar^n \hat{\Delta} . \tag{3.41}$$

It leads to the new vertex functional

$$\Gamma_{(n)} = \Gamma_{(n-1)} - \hbar^n \hat{\Delta} + O(\hbar^{n+1}) . \tag{3.42}$$

Thus we get

$$\mathcal{W}\Gamma_{(n)} = O(\hbar^{n+1}) , \tag{3.43}$$

which is the next order Ward identity we wanted to prove.

*Remarks.*

1. The coefficients $r^i$ are not free. They are fixed in terms of the parameters of the action $S_{(n-1)}$, which was assumed to be known.

2. It is clear that, at the order $n$ we are considering, we are still free to add to the action any invariant counterterm of dimension bounded by four. These counterterms are given by the three invariant pieces of the classical action (3.32):

$$\int d^4x \; \partial\bar{\varphi}\partial\varphi , \quad \int d^4x \; \bar{\varphi}\varphi , \quad \int d^4x \; (\bar{\varphi}\varphi)^2 . \tag{3.44}$$

The corresponding coefficients have to be fixed by three normalization conditions quite analogous to the conditions (2.56) for the scalar $\varphi^4$ theory:

$$\begin{aligned}
\tilde{\Gamma}_{\varphi\bar{\varphi}}(p^2)\Big|_{p^2=-m^2} &= 0 , \\
\frac{d}{dp^2}\tilde{\Gamma}_{\varphi\bar{\varphi}}(p^2)\Big|_{p^2=\mu^2} &= 1 , \\
\tilde{\Gamma}_{\varphi\varphi\bar{\varphi}\bar{\varphi}}(p_1 \cdots p_4)\Big|_{p_i^2=\mu^2, \; (p_i+p_j)^2=\frac{4}{3}\mu^2 \; (i\neq j)} &= g .
\end{aligned} \tag{3.45}$$

3. The free invariant counterterms are in one to one correspondence with the terms of the classical action. This means that they correspond to a renormalization of the initial parameters of the theory. In such a case one speaks of a *multiplicatively renormalizable* theory. In more physical terms, this property expresses the *stability* of the theory upon quantum fluctuations.

## 3.4.2 The Callan–Symanzik Equation

In the classical theory the scale invariance is broken by the mass term of the action. The scale being fixed by the value of the mass $m$, a scale transformation is generated by the operator $m\partial_m$. The application of this operator to the action (3.32) yields

$$m\partial_m S = 2m^2 \int d^4x \; \bar{\varphi}\varphi . \tag{3.46}$$

The right–hand side is a (classical) insertion of dimension 2. Two changes occur in the quantum theory. First, in the U(1) model considered up to now, there is a second dimensionful parameter, the normalization mass $\mu$ (see (3.45)), so that the scale operator is now

$$m\partial_m + \mu\partial_\mu \ . \tag{3.47}$$

Secondly the insertion in the right–hand side is not necessarily of dimension 2: according to the quantum action principle (3.28) one has

$$(m\partial_m + \mu\partial_\mu)\Gamma = \Delta \cdot \Gamma \ , \quad \dim(\Delta) \leq 4 \ . \tag{3.48}$$

Moreover, use of the Ward identity (3.36) and of the fact the scale operator (3.47) commutes with the operator (3.35),

$$[m\partial_m + \mu\partial_\mu \, , \mathcal{W}] = 0 \ , \tag{3.49}$$

implies that $\Delta \cdot \Gamma$ is a symmetric insertion, i.e.,

$$\mathcal{W}\Delta \cdot \Gamma = 0 \ . \tag{3.50}$$

Thus, according to Prop. 3.1, a basis in which one can expand this insertion is provided by any quantum extension of the classical insertions (3.44). An equivalent basis is provided by

$$N\Gamma := \int d^4x \ \left(\varphi\frac{\delta}{\delta\varphi} + \bar\varphi\frac{\delta}{\delta\bar\varphi}\right)\Gamma \ , \quad \Delta_m \cdot \Gamma \ , \quad \partial_g\Gamma \ . \tag{3.51}$$

The first and the third objects are clearly quantum extensions of the first and third elements of the basis (3.44) – up to numerical factors – due to the quantum action principle. The second term of (3.51) is a quantum extension of dimension 2 of the mass insertion $\int \bar\varphi\varphi$. Its normalization is fixed by the condition

$$\Delta_m \cdot \tilde\Gamma_{\varphi\bar\varphi}\Big|_{p^2=\mu^2} = 1 \ . \tag{3.52}$$

The expansion of (3.48) in the basis (3.51) yields the Callan–Symanzik equation [57, 62]

$$(m\partial_m + \mu\partial_\mu + \beta\partial_g - \gamma N)\,\Gamma = 2m^2(1+\delta)\Delta_m \cdot \Gamma \ . \tag{3.53}$$

The coefficients $\beta$, $\gamma$ and $\delta$ are calculable – for example through the normalization conditions – functions of the parameters of the theory. $\beta$ expresses the renormalization of the coupling constant following a change of scale, whereas $\gamma$ expresses the renormalization of the field amplitude.

*Remark.* Use of the normalization conditions (3.45) explicitly yields

$$\delta = \frac{1}{2m^2}\left[m\partial_m\tilde\Gamma_{\varphi\bar\varphi}\left[\Delta_m \cdot \tilde\Gamma_{\varphi\bar\varphi}\right]^{-1}\right]_{p^2=-m^2} - 1 \ ,$$

$$\gamma = \left[\tfrac{1}{2}\mu\partial_\mu\frac{d}{dp^2}\tilde\Gamma_{\varphi\bar\varphi} - m^2(1+\delta)\frac{d}{dp^2}\Delta_m \cdot \tilde\Gamma_{\varphi\bar\varphi}\right]_{p^2=\mu^2} \ , \tag{3.54}$$

$$\beta = 4\gamma g + \left[-\mu\partial_\mu\tilde\Gamma_{\varphi\varphi\bar\varphi\bar\varphi} + 2m^2(1+\delta)\Delta_m \cdot \tilde\Gamma_{\varphi\varphi\bar\varphi\bar\varphi}\right]_{p_i^2=\mu^2, \ (p_i+p_j)^2=\frac{4}{3}\mu^2 \ (i\neq j)} \ .$$

At the order $\hbar$ this yields

$$
\begin{aligned}
\delta^{(1)} &= \frac{1}{2m^2}\left[m\partial_m \tilde{\Gamma}^{(1)}_{\varphi\bar\varphi} - 2m^2\Delta_m \cdot \tilde{\Gamma}^{(1)}_{\varphi\bar\varphi}\right]_{p^2=-m^2} , \\
\gamma^{(1)} &= \left[\tfrac{1}{2}\mu\partial_\mu \frac{d}{dp^2}\tilde{\Gamma}^{(1)}_{\varphi\bar\varphi} - m^2\frac{d}{dp^2}\Delta_m \cdot \tilde{\Gamma}^{(1)}_{\varphi\bar\varphi}\right]_{p^2=\mu^2} , \\
\beta^{(1)} &= 4\gamma^{(1)}g + \left[-\mu\partial_\mu \tilde{\Gamma}^{(1)}_{\varphi\varphi\bar\varphi\bar\varphi} + 2m^2\Delta_m \cdot \tilde{\Gamma}^{(1)}_{\varphi\varphi\bar\varphi\bar\varphi}\right]_{p_i^2=\mu^2,\ (p_i+p_j)^2=\frac{4}{3}\mu^2\ (i\neq j)}
\end{aligned}
\tag{3.55}
$$

where the superscript (1) means one–loop graph contributions.

### 3.4.3  Renormalization Group Equation

Whereas the Callan–Symanzik equation describes the response of the theory to a change of scale, as we have seen in Sect. 3.4.2, the renormalization group equation will describe the response to a change of the normalization conditions – c.f. (2.56) or (3.45) – defining the parametrization of the theory. These normalization conditions being labelled by a normalization mass $\mu$, such a change is generated by the differential operator $\mu\partial/\partial\mu$. Following the same strategy as for the Callan–Symanzik equation, we expand the insertion

$$
\mu\frac{\partial}{\partial\mu}\Gamma = \Delta\cdot\Gamma
$$

in a basis given now by the three insertions generated by the operators $m\partial/\partial m$, $\partial/\partial g$ and $N$, the latter being the "counting operator" defined in (3.51). We thus obtain the renormalization group equation

$$
\left(\mu\partial_\mu + \hat{\beta}_m m\partial_m + \hat{\beta}_g\partial_g - \hat{\gamma}N\right)\Gamma = 0 ,
\tag{3.56}
$$

which clearly expresses the renormalization of the mass, of the coupling constant and of the field amplitude caused by a change of the normalization mass $\mu$.

However if, as it is the case here (see (3.45)), the mass $m$ is defined as the position of the propagator pole, then $m$ is not renormalized, i.e., the following renormalization group equation holds:

$$
\left(\mu\partial_\mu + \hat{\beta}_g\partial_g - \hat{\gamma}N\right)\Gamma = 0 .
\tag{3.57}
$$

The proof of this is quite simple. We first write (3.56) for the two–point vertex function (i.e., for the inverse propagator), written as

$$
\tilde{\Gamma}_{\varphi\bar\varphi}(p^2) = \lambda(p^2)\left(p^2 + m^2\right)
$$

due to the first of the normalization conditions (3.45):

$$
\left(\mu\partial_\mu + \hat{\beta}_m m\partial_m + \hat{\beta}_g\partial_g - 2\hat{\gamma}\right)\lambda(p^2)\left(p^2 + m^2\right) = 0 .
$$

Setting then $p^2 + m^2 = 0$, we are left with the equation

$$
2\hat{\beta}_m\lambda(-m^2)m^2 = 0 ,
$$

which implies the vanishing of $\hat{\beta}_m$.

*Remark.* In the massless case the renormalization group equation coincides with the Callan–Symanzik equation, the right–hand side of the latter vanishing as $m \to 0$. This is physically obvious: the normalization mass parameter $\mu$ being then the only dimensionful parameter, fixes the scale. In the general case the two equations are quite different objects. In particular the Callan–Symanzik $\beta$–function differs from the renormalization group $\hat{\beta}$–function.

## 3.5 A O($N$) Scalar Model

We consider now a slight generalization of the four–dimensional model of the preceding section, changing from the Abelian group U(1) to the nonabelian group O($N$) ($N \geq 3$). The fields $(\varphi_i(x), i = 1, \cdots, N)$ are scalar again, and belong to a $N$–plet transforming (infinitesimally) under O($N$) as:

$$\delta\varphi_i = i\varepsilon_a R_i^{aj}\varphi_j , \tag{3.58}$$

where $\varepsilon_a$ are constant infinitesimal parameters and the $\frac{1}{2}N(N-1)$ Hermitian matrices $R^a$ are the generators of the group, obeying the Lie algebra commutation relations

$$[R^a, R^b] = if^{abc}R^c . \tag{3.59}$$

### 3.5.1 Ward Identities

The corresponding functional variational operators (Ward operators) are defined by

$$\mathcal{W}^a = -\int d^4x \; R_i^{aj}\varphi_j \frac{\delta}{\delta\varphi_i} . \tag{3.60}$$

They obey the commutation rules of the Lie algebra:

$$[\mathcal{W}^a, \mathcal{W}^b] = if^{abc}\mathcal{W}^c . \tag{3.61}$$

The O($N$) invariance of the classical theory is expressed by the classical Ward identities

$$\mathcal{W}^a S = 0 . \tag{3.62}$$

The most general power–counting renormalizable classical action $S$ solution of these Ward identities is

$$S = \int d^4x \; \left(\frac{1}{2}\partial\varphi^i\partial\varphi_i + \frac{m^2}{2}\varphi^i\varphi_i + \frac{g}{4}(\varphi^i\varphi_i)^2\right) . \tag{3.63}$$

The same recursive procedure as for the U(1) model can be applied to show that the Ward identities can be maintained to all orders in $\hbar$: the problem being assumed to be solved up to and including the order $n-1$, we can write, using the quantum action principle (3.27),

$$\mathcal{W}^a\Gamma = \hbar^n\Delta^a \cdot \Gamma = \hbar^n\Delta^a + O(\hbar^{n+1}) , \tag{3.64}$$

where the $\Delta^a$'s are $\frac{1}{2}N(N-1)$ integrated local insertions of dimension less or equal to four. If these insertions were of the form

$$\Delta^a = \mathcal{W}^a \hat{\Delta} \ , \tag{3.65}$$

with $\hat{\Delta}$ an integrated local insertion of dimension bounded by four, then a redefinition of the action $S \to S - \hbar^n \hat{\Delta}$, hence of the vertex functional $\Gamma \to \Gamma - \hbar^n \hat{\Delta} + O(\hbar^{n+1})$, would implement the Ward identities to order $n$ (cf. (3.41)-(3.43)). In order to check if (3.65) actually holds, we can exploit the algebraic structure displayed in the commutation rules (3.61). Indeed applying these Ward operators identities to the vertex functional:

$$\left( [\mathcal{W}^a, \mathcal{W}^b] - if^{abc}\mathcal{W}^c \right) \Gamma = 0 \ ,$$

making use of (3.64) and selecting the resulting terms of order $n$, one obtains the *consistency condition*

$$\mathcal{W}^a \Delta^b - \mathcal{W}^b \Delta^a - if^{abc}\Delta^c = 0 \ . \tag{3.66}$$

### 3.5.2 Solution of the Consistency Condition

Solving constraints such as (3.66) is technically known as a problem of Lie algebra cohomology. Instead of the O($N$) symmetry we may consider the slightly more general case of a semi–simple Lie group $G$, with the fields belonging to any finite representation. We shall follow the argument of [17] (see also [63]). Let us denote by $X^a$, with

$$\left[ X^a, X^b \right] = if^{abc} X^c \ ,$$

the generators of $G$ in a basis where the invariant quadratic form – the Killing form – is equal to $\delta^{ab}$ and the structure constants are completely antisymmetric. The insertion $\Delta^a$ being an integrated local field polynomial of bounded dimension, it belongs to a finite representation, which can be decomposed into irreducible representations. Its expansion into these irreducible representations reads

$$\Delta^a = \Delta_{\text{inv}}^a + \sum_R \Delta_R^a \ ,$$

where $\Delta_{\text{inv}}^a$ belongs to the trivial representation and the sum extends over the nontrivial irreducible representations $R$. The consistency condition (3.66) then decomposes into the equations

$$f^{abc}\Delta_{\text{inv}}^c = 0 \ ,$$
$$\mathcal{W}^a \Delta_R^b - \mathcal{W}^b \Delta_R^a - if^{abc}\Delta_R^c = 0 \ . \tag{3.67}$$

The first one gives

$$\Delta_{\text{inv}}^c = 0 \ , \tag{3.68}$$

due to the assumption of a semisimple group. The application of the Ward operator $\mathcal{W}^a$ to the second equation yields

$$c_R \Delta_R^b - \mathcal{W}^b \mathcal{W}^a \Delta_R^a = 0 \ , \tag{3.69}$$

where we have used the Lie algebra commutation relations and the antisymmetry of the structure constants. The real number $c_R$ is the eigenvalue of the quadratic Casimir operator whose action on the field polynomials is given by the functional operator $\mathcal{W}^a \mathcal{W}^a$. Since this eigenvalue is different from zero for any nontrivial irreducible representation, (3.69) implies that each $\Delta_R^a$ is the $\mathcal{W}^a$–variation of some integrated local field polynomial. Taking the result (3.68) into account we arrive to the conclusion that the whole $\Delta^a$ is such a variation, i.e., that the cohomology expressed by (3.66) is trivial[4]. Having thus shown that the general solution of the consistency conditions has the form (3.65) we conclude that there is no anomaly: the Ward identities

$$\mathcal{W}^a \Gamma = 0 \tag{3.70}$$

can be implemented to all orders.

*Remark.* In the context of power–counting renormalizable theories (see Subsect. 2.2.5), the renormalizability of a rigid symmetry, i.e., the possibility of implementing it to all orders, can be generalized [17] to the case of any Lie group $G$ which is the product of a semisimple group and of U(1) factors, and also in the presence of a breaking of arbitrary covariance. A typical example of the latter is provided by the breaking produced by a noninvariant mass term transforming in some representation of $G$.

### 3.5.3 The Callan–Symanzik Equation

As usually arbitrary invariant counterterms can be absorbed at each order. They amount to a finite renormalization of the parameters of the theory and they can be fixed by normalization conditions at a normalization scale $\mu$. Corresponding to this there exists also a Callan–Symanzik equation which, for the $O(N)$ model, writes

$$(m\partial_m + \mu\partial_\mu + \beta\partial_g - \gamma N)\,\Gamma = m^2(1+\delta)\Delta_m \cdot \Gamma \,, \tag{3.71}$$

where $N$ is the counting operator

$$N = \int d^4x \sum_i \varphi_i \frac{\delta}{\delta\varphi_i} \,, \tag{3.72}$$

and $\Delta_m$ is a symmetric insertion – i.e., $\mathcal{W}^a \Delta_m \cdot \Gamma = 0$ – which is a quantum extension of dimension 2 of the mass insertion $\int \varphi^i \varphi_i$.

*Remark.* The differential operators appearing in the Callan–Symanzik equation are *symmetric*, i.e., they commute with the Ward identity operators. In general a symmetric operator $\nabla$,

$$[\mathcal{W}^a, \nabla] = 0 \,, \tag{3.73}$$

generates through the quantum action principle a symmetric insertion $\Delta_\nabla$:

$$\nabla\Gamma = \Delta_\nabla \cdot \Gamma \,, \qquad \mathcal{W}^a \Delta_\nabla \cdot \Gamma = 0 \,. \tag{3.74}$$

---

[4]This general result, which states that the first cohomology of a semisimple algebra with values in a finite representation is trivial, is known as Whitehead's Lemma [64].

## 3.6 Basis of Symmetric Insertions

Let us consider a theory defined by a symmetry described by the set of Ward identities[5]

$$\mathcal{W}_a \Gamma = 0, \quad a = 1, 2, \cdots , \tag{3.75}$$

such as, for example, the $O(N)$ model of Sect. 3.5. Quantum insertions $\Delta^p \cdot \Gamma$ are defined as symmetric by the property

$$\mathcal{W}_a \Delta^p \cdot \Gamma = 0 . \tag{3.76}$$

Their classical approximations evidently obey the classical constraints

$$\mathcal{W}_a \Delta^p_{\text{class}} = 0 . \tag{3.77}$$

Prop. 3.1 generalizes itself to the case of such symmetric insertions:

**Proposition 3.2** *Let*

$$\left\{ \Delta^p \cdot \Gamma = \Delta^p_{\text{class}} + O(\hbar) \mid p = 1, 2, \cdots ; \dim (\Delta^p) \leq \bar{d} \right\} \tag{3.78}$$

*be a set of symmetric insertions, obeying the invariance conditions (3.76) and such that their classical approximations $\Delta^p_{\text{class}}$ form a basis for the symmetric classical insertions of dimension bounded by $\bar{d}$. Then the set (3.78) is a basis for the quantum symmetric insertions of dimension bounded by $\bar{d}$.*

The proof is a paraphrase of that of Prop. 3.1 and will not be repeated.

## 3.7 Summary of the Quantum Action Principle

For the convenience of the reader, let us summarize the main content of the quantum action principle, applied here to the vertex functional.

Let $\Gamma(\phi_a, \rho_i, \lambda)$ be the vertex functional corresponding to a power–counting renormalizable theory with a classical action given by

$$S = \int d^D x \mathcal{L}(\phi_a, \rho_i, \lambda) , \tag{3.79}$$

where $\phi_a$ denotes a set of fields with dimensions $d_a$, $\rho_i$ a set of external fields coupled to field polynomials $Q^i$ of dimensions $d_{Q^i}$, and $\lambda$ the set of the parameters (masses, coupling constants, renormalization point). Then:

i) **Equations of Motion:**

$$\frac{\delta \Gamma}{\delta \phi_a(x)} = \Delta^a(x) \cdot \Gamma . \tag{3.80}$$

$\Delta^a(x)$ is a local insertion of dimension bounded by $(D - d_a)$. Its lowest order is given by

$$\Delta^a \cdot \Gamma = \frac{\delta S}{\delta \phi_a} + O(\hbar) . \tag{3.81}$$

---

[5]We restrict ourselves here to the case where the Ward identity operators $\mathcal{W}_a$ act linearly on the vertex functional $\Gamma$.

ii) **Nonlinear Variations of the Fields:**

$$\frac{\delta\Gamma}{\delta\rho_i(x)}\frac{\delta\Gamma}{\delta\phi_a(x)} = \Delta^{ai}(x)\cdot\Gamma \ . \tag{3.82}$$

$\Delta^{ai}(x)$ is a local insertion of dimension bounded by $(D - d_a + d_{Q^i})$ and, to lowest order:

$$\Delta^{ai}\cdot\Gamma = \frac{\delta S}{\delta\rho_i}\frac{\delta S}{\delta\phi_a} + O(\hbar) \ . \tag{3.83}$$

iii) **Variation With Respect to a Parameter:**

$$\frac{\partial\Gamma}{\partial\lambda} = \Delta\cdot\Gamma \ . \tag{3.84}$$

$\Delta$ is a space–time integrated insertion of dimension bounded by $D$. At lowest order the latter reads

$$\Delta\cdot\Gamma = \frac{\partial S}{\partial\lambda} + O(\hbar) \ . \tag{3.85}$$

*Remark.* In the case of a purely massless theory, the dimension bounds given above become equalities.

# 4. Yang–Mills Gauge Theories

In this chapter we shall deal with Ward identities and with their renormalization in the case of a *local invariance*. We shall restrict the discussion here to Yang–Mills theories in four–dimensional space–time. Beyond their obvious physical relevance [2], these theories have played an eminent historical role. It was indeed in the search for a rigorous renormalization scheme for these theories that Becchi, Rouet and Stora [15, 16] in 1974 discovered BRS invariance and demonstrated its role in guarantying the unitarity of the $S$–matrix and the gauge independence of the physical observables, on the one hand, and set up the general iterative construction needed in the absence of an invariant regularization. Precisely the latter approach is called "algebraic renormalization" in the present book.

For illustrative purposes, we shall consider the model of a Yang–Mills gauge field corresponding to a simple Lie group, coupled to a multiplet of left–handed fermions. After having established its properties in the tree graph approximation, in particular its BRS invariance, we shall undertake the perturbative construction of the corresponding quantum theory and classify the possible obstructions to this construction: the gauge anomalies. The latter, as well as the invariant counterterms corresponding to a redefinition of the physical parameters, are characterized by the cohomology of the BRS operator.

Let us emphasize that the preservation of the BRS symmetry to all orders is the basic ingredient in the proof of the unitarity of the scattering matrix and generally of the construction of gauge invariant operators. We shall not give here the proof of the unitarity of the scattering operator, for which the reader may look at the original literature quoted above and at [65, 66] for more recent accounts. On the other hand, we shall explain how the BRS invariance can be used in order to define gauge invariant observables.

## 4.1 Tree Approximation, Gauge Fixing and BRS Invariance

The gauge theory in four–dimensional space–time considered in this chapter is based on a simple compact Lie group. The gauge fields are Lie algebra valued – they belong to the adjoint representation:

$$A_\mu(x) = A_\mu^a(x)\tau_a .$$ 

(4.1)

The matrices $\tau$ are the generators of the group and obey

$$[\tau_a, \tau_b] = i f_{abc} \tau_c , \qquad \mathrm{Tr}\, \tau_a \tau_b = \delta_{ab} ,$$ 

(4.2)

where Tr is the normalized trace and defines the scalar product.

The matter fields consist of a left–handed fermion multiplet

$$\psi(x) = \psi_{\text{left}}(x) , \qquad \text{with} \quad \frac{1}{2}(1 - \gamma_5)\psi_{\text{left}}(x) = \psi_{\text{left}}(x) , \tag{4.3}$$

belonging to some finite representation of the gauge group.

The infinitesimal gauge transformations are defined as

$$\delta A_\mu(x) = \partial_\mu \omega(x) + i[\omega(x), A_\mu(x)] = D_\mu \omega(x) ,$$
$$\delta \psi(x) = i\omega^a(x) T_a \psi(x) = i\omega(x)\psi(x) . \tag{4.4}$$

In the first line $\omega = \omega^a \tau_a$, in the second line $\omega = \omega^a T_a$, where the matrices $T_a$ represent the generators in the fermion representation.

*Remark.* The infinitesimal transformations (4.4) are obtained from the finite gauge transformations

$$A_\mu^{\mathcal{U}}(x) = \mathcal{U} A_\mu(x)\mathcal{U}^{-1} + i\mathcal{U}\partial_\mu\mathcal{U}^{-1} ,$$
$$\psi^{\mathcal{U}}(x) = \mathcal{U}\psi(x) , \tag{4.5}$$
$$\text{with} \quad \mathcal{U} = e^{i\omega(x)} ,$$

by keeping the linear terms of their power expansion in $\omega$.

The Yang–Mills field strength and the covariant derivative read

$$F_{\mu\nu} = \partial_\mu A_\nu - \partial_\nu A_\mu - i[A_\mu, A_\nu] ,$$
$$D_\mu \psi = (\partial_\mu - iA_\mu^a T_a)\psi . \tag{4.6}$$

The most general gauge invariant and power–counting renormalizable[1] action is

$$S_{\text{inv}} = \int d^4x \left( -\frac{1}{4g^2} \text{Tr} \, F^{\mu\nu} F_{\mu\nu} + i\bar{\psi}\slashed{D}\psi \right) , \tag{4.7}$$

where

$$\slashed{D} = \gamma^\mu D_\mu .$$

Note that the choice of a chiral representation (see (4.3)) for the fermions forbids any mass term.

The gauge fixing is done in the BRS way [16, 65]. One introduces Faddeev–Popov ghost and antighost fields $c(x)$ and $\bar{c}(x)$, and the Lautrup–Nakanishi [67] field $B(x)$. The latter plays the role of a Lagrange multiplier for the gauge condition. These three fields are Lie algebra valued: $c(x) = c^a(x)\tau_a$, etc. The components of $c$ and $\bar{c}$ are anticommuting. One chooses a covariant linear gauge condition, which one can express by the functional identity

$$\frac{\delta S}{\delta B} = \partial^\mu A_\mu + \alpha B . \tag{4.8}$$

---

[1]According to the rules established in Sect. 2.2 this means that the dimension of the terms in the action must not exceed 4, the dimensions of the various fields being given in Table 4.1.

This means that the Lagrange multiplier dependent terms of the action are prescribed to be:

$$\text{Tr} \int d^4x \left( B\partial A + \frac{1}{2}\alpha B^2 \right) .$$

The special case with $\alpha = 0$ is called the Landau gauge.

The gauge transformations are replaced by the BRS transformations

$$sA_\mu = \partial_\mu c + i[c, A_\mu] = D_\mu c ,$$

$$s\psi = ic^a T_a \psi , \qquad\qquad\qquad s\bar\psi = i\bar\psi T_a c^a ,$$

$$sc = ic^2 \qquad (sc^a = -\tfrac{1}{2}f_{abc}c^b c^c) ,$$ \hfill (4.9)

$$s\bar c = B , \qquad\qquad\qquad sB = 0 ,$$

which are nilpotent:

$$s^2 = 0 . \qquad\qquad (4.10)$$

The BRS transformations of some of the fields being nonlinear, one has to introduce BRS invariant external fields $\rho^\mu$, $Y$ and $\sigma$ coupled to them. This is done by adding to the action the terms

$$S_{\text{ext}} = \int d^4x \left( \text{Tr}\left( \rho^\mu s A_\mu \right) + \bar Y s\psi - s\bar\psi Y + \text{Tr}\left( \sigma sc \right) \right) . \qquad (4.11)$$

$\rho^\mu$ and $\sigma$ are Lie algebra valued, whereas the right–handed spinor field $Y$ transforms in the same representation as $\psi$. BRS invariance can thus be expressed by a functional identity, the *Slavnov–Taylor identity* [15, 16, 68, 69]:

$$\mathcal{S}(S) = 0 , \qquad\qquad (4.12)$$

where the nonlinear Slavnov–Taylor operator is given, for any functional $\mathcal{F}$, by[2]

$$\mathcal{S}(\mathcal{F}) = \int d^4x \left( \text{Tr}\, \frac{\delta\mathcal{F}}{\delta\rho^\mu}\frac{\delta\mathcal{F}}{\delta A_\mu} + \frac{\delta\mathcal{F}}{\delta\bar Y}\frac{\delta\mathcal{F}}{\delta\psi} - \frac{\delta\mathcal{F}}{\delta Y}\frac{\delta\mathcal{F}}{\delta\bar\psi} + \text{Tr}\, \frac{\delta\mathcal{F}}{\delta\sigma}\frac{\delta\mathcal{F}}{\delta c} + \text{Tr}\, B\frac{\delta\mathcal{F}}{\delta\bar c} \right) . \qquad (4.13)$$

The most general power–counting renormalizable, BRS invariant action obeying the gauge condition (4.8) and the Slavnov–Taylor identity is

$$S = S_{\text{inv}} + S_{\text{gf}} + S_{\text{ext}} , \qquad\qquad (4.14)$$

with $S_{\text{inv}}$, $S_{\text{ext}}$ given in (4.7), (4.11), and with the gauge fixing action $S_{\text{gf}}$ given by

$$\begin{aligned}
S_{\text{gf}} &= s\text{Tr} \int d^4x \left( \bar c\partial^\mu A_\mu + \frac{\alpha}{2}\bar c B \right) \\
&= \text{Tr} \int d^4x \left( B\partial^\mu A_\mu + \frac{\alpha}{2}B^2 - \bar c\partial^\mu(\partial_\mu c + i[c, A_\mu]) \right) .
\end{aligned} \qquad (4.15)$$

The action (4.14) preserves the *ghost number*. The values of the ghost number are displayed together with the dimensions in Table 4.1. The statistics is defined as follows: the fields of integer spin and odd ghost number as well as the fields of half integer spin and even ghost number are anticommuting; the other fields commute with the formers and among themselves.

---

[2] $A_\mu$ and $c$ are chosen real, $\bar c$, $\rho^\mu$ and $\sigma$ imaginary. Moreover $\bar\psi = \psi^+\gamma^0$, $\bar Y = Y^+\gamma^0$. With this choice the action as well as the Slavnov–Taylor operator are real.

**Table 4.1.** Ghost numbers and dimensions.

| | $s$ | $A_\mu$ | $\psi$ | $c$ | $\bar{c}$ | $B$ | $\rho^\mu$ | $Y$ | $\sigma$ |
|---|---|---|---|---|---|---|---|---|---|
| Ghost number | 1 | 0 | 0 | 1 | −1 | 0 | −1 | −1 | −2 |
| Dimension | 0 | 1 | 3/2 | 0 | 2 | 2 | 3 | 5/2 | 4 |

*Remarks.*

1. The gauge choice defined by the gauge fixing condition (4.8) belongs to the class of the linear, covariant gauges discussed by 't Hooft [10]. Besides of them there is a wide class of linear, noncovariant gauges [70, 71], for example the axial gauge [72] and the light–cone gauge [73][3]. Finally nonlinear covariant gauges were also considered [11, 74, 75, 76]. The right–hand side of (4.8) being in this case a polynomial nonlinear in the fields, it could require the introduction of an additional external field in order to control its quantum extension. As an example, the renormalization of quantum electrodynamics in a nonlinear gauge was studied in [77].

2. The Lagrange multiplier field $B$ could be eliminated by using its equation of motion (4.8), which does not involve the derivatives of $B$. In fact the resulting action would be the one used in all the early papers on the subject [16, 69, 61] and would be completely equivalent to the action where $B$ is not eliminated. One disadvantage of working without $B$ is that BRS nilpotency would hold only modulo the field equation of the ghost $c$ (given by the equation (4.20) below). Moreover, as we shall see in Sect. 4.5 the use of the Lagrange multiplier field allows to control in a simpler way the dependence on the gauge parameter $\alpha$. This is due to the important property that the derivative of the action with respect to $\alpha$ is an exact BRS variation:

$$\frac{\partial S}{\partial \alpha} = s \mathrm{Tr} \int d^4x \, \frac{\bar{c}B}{2} \, . \tag{4.16}$$

## 4.2 Renormalization: The Consistency Condition

We want to see if it is possible to extend, perturbatively, the classical theory described by the action (4.14) to a quantum theory described by a vertex functional $\Gamma(A, \psi, c, \bar{c}, B, \rho, Y, \sigma)$ obeying the same Slavnov–Taylor identity (4.12) as the classical action.

As in the classical case we shall define the theory by the *gauge fixing* condition (cf. (4.15))

---

[3]These noncovariant gauges are affected by specific problems [73], like the difficulty to properly define a causal propagator, or the appearance of divergences corresponding to nonlocal counterterms.

$$\frac{\delta\Gamma}{\delta B} = \partial^{\mu}A_{\mu} + \alpha B \ , \tag{4.17}$$

and by the *Slavnov–Taylor identity*

$$\mathcal{S}(\Gamma) = 0 \ , \tag{4.18}$$

which in terms of the Green functional $Z$ reads

$$SZ = \int d^4x \ \left( -\mathrm{Tr} \ J_A^{\mu}\frac{\delta}{\delta\rho^{\mu}} + J_{\psi}\frac{\delta}{\delta Y} + J_{\bar{\psi}}\frac{\delta}{\delta\bar{Y}} + \mathrm{Tr} \ J_c\frac{\delta}{\delta\sigma} + \mathrm{Tr} \ J_{\bar{c}}\frac{\delta}{\delta J_B} \right) Z = 0 \ . \tag{4.19}$$

Let us begin by showing that (4.17) is not affected by the radiative corrections. We shall use the method introduced in the subsections 3.4.1 and 3.5.1 in the case of the Ward identities associated with a rigid symmetry. Assuming (4.17) to hold below the order $n$ in $\hbar$, the most general breaking compatible with the power–counting constraints reads

$$\frac{\delta\Gamma}{\delta B^a} = \partial^{\mu}A^a_{\mu} + \alpha B^a + \hbar^n\Delta^a + O(\hbar^{n+1}) \ ,$$

with

$$\Delta^a = F^a(A, c, \bar{c}) + \omega_{ab}B^b \ ,$$

where $F^a$ is a local polynomial in $A$, $c$ and $\bar{c}$ of dimension two, and the $\omega_{ab}$ are numbers. From $[\delta/\delta B^a(x), \delta/\delta B^b(y)] = 0$ follows the consistency condition

$$\frac{\delta}{\delta B^a(x)}\Delta^b(y) - \frac{\delta}{\delta B^b(y)}\Delta^a(x) = 0 \ ,$$

which implies $\omega_{ab} = \omega_{ba}$. This is an integrability condition showing that the breaking can be written as a functional derivative with respect to $B$:

$$\Delta^a(x) = \frac{\delta}{\delta B^a(x)}\int d^4y \ \left( B^aF^a + \frac{1}{2}\omega_{ab}B^aB^b \right)(y) =: \frac{\delta}{\delta B^a(x)}\tilde{\Delta}$$

The absorption of $-\tilde{\Delta}$ as a counterterm at the order $n$ then assures the validity of (4.17) at this order. This ends the recursive proof of the renormalizability of the gauge condition.

*Remark.* The right–hand side of (4.17) being linear in the fields is well defined as it stands without the need of renormalization – recall that these fields belong to the set of the $C^{\infty}$ "classical fields", arguments of the vertex functional.

A further identity, the *ghost equation*

$$\mathcal{G}\Gamma = 0 \ , \qquad \text{with} \ \ \mathcal{G} := \frac{\delta}{\delta\bar{c}} + \partial^{\mu}\frac{\delta}{\delta\rho^{\mu}} \ . \tag{4.20}$$

follows from the gauge fixing condition and from the Slavnov–Taylor identity. This fact is a consequence of the identity, true for any functional $\mathcal{F}$,

$$\frac{\delta}{\delta B}\mathcal{S}(\mathcal{F}) - \mathcal{S}_{\mathcal{F}}\left( \frac{\delta\mathcal{F}}{\delta B} - \partial^{\mu}A_{\mu} \right) = \mathcal{G}\mathcal{F} \ , \tag{4.21}$$

where we have introduced the linear, functional dependent, operator

$$
\begin{aligned}
\mathcal{S}_{\mathcal{F}} = \int d^4x \bigg( &\mathrm{Tr}\, \frac{\delta \mathcal{F}}{\delta \rho^\mu}\frac{\delta}{\delta A_\mu} + \mathrm{Tr}\, \frac{\delta \mathcal{F}}{\delta A_\mu}\frac{\delta}{\delta \rho^\mu} + \frac{\delta \mathcal{F}}{\delta \bar{Y}}\frac{\delta}{\delta \psi} + \frac{\delta \mathcal{F}}{\delta \psi}\frac{\delta}{\delta \bar{Y}} \\
&- \frac{\delta \mathcal{F}}{\delta Y}\frac{\delta}{\delta \bar{\psi}} - \frac{\delta \mathcal{F}}{\delta \bar{\psi}}\frac{\delta}{\delta Y} + \mathrm{Tr}\, \frac{\delta \mathcal{F}}{\delta \sigma}\frac{\delta}{\delta c} + \mathrm{Tr}\, \frac{\delta \mathcal{F}}{\delta c}\frac{\delta}{\delta \sigma} + \mathrm{Tr}\, B\frac{\delta}{\delta \bar{c}} \bigg) .
\end{aligned}
\tag{4.22}
$$

$\mathcal{S}_{\mathcal{F}}$ is called the *linearized Slavnov–Taylor operator*. For further use let us write the "anticommutation relation"

$$
\mathcal{G}\mathcal{S}(\mathcal{F}) + \mathcal{S}_{\mathcal{F}}\mathcal{G}\mathcal{F} = 0 .
\tag{4.23}
$$

The proof of the renormalizability of the ghost equation is identical to the the proof given above for the gauge condition (4.17). Indeed, after a change of variables $\rho \to \hat{\rho}$ defined by

$$
\hat{\rho}^\mu = \rho^\mu + \partial^\mu \bar{c} ,
\tag{4.24}
$$

all the other fields remaining unchanged, the ghost equation (4.20) takes the form

$$
\frac{\delta \Gamma}{\delta \bar{c}} = 0 ,
$$

which, as the left–hand side of the gauge condition (4.17), is a simple functional derivative.

Having thus proved the implementability of the gauge condition and of the ghost equation to all orders, we assume them to hold from now on.

*Remark.* It follows from the gauge fixing condition and from the ghost equation that the vertex functional can be written as

$$
\Gamma(A,\psi,c,\bar{c},B,\rho,Y,\sigma) = \hat{\Gamma}(A,\psi,c,\hat{\rho},Y,\sigma) + \mathrm{Tr}\int d^4x \left( B\partial A + \frac{\alpha}{2}B^2 \right) ,
\tag{4.25}
$$

where $\hat{\Gamma}$ is independent of the Lagrange multiplier field $B$, and depends on the ghost $\bar{c}$ and on the external field $\rho$ only through the combination $\hat{\rho}$ given by (4.24).

The classical action, of course, obeys the same decomposition

$$
S(A,\psi,c,\bar{c},B,\rho,Y,\sigma) = \hat{S}(A,\psi,c,\hat{\rho},Y,\sigma) + \mathrm{Tr}\int d^4x \left( B\partial A + \frac{\alpha}{2}B^2 \right) .
\tag{4.26}
$$

It remains to prove the validity of the Slavnov–Taylor identity (4.18). The first task is to express in a functional way the nilpotency of the BRS operation. This property is contained in the two identities

$$
\begin{aligned}
\mathcal{S}_{\mathcal{F}}\mathcal{S}(\mathcal{F}) &= 0 , \quad \forall\, \mathcal{F} , \\
\mathcal{S}_{\mathcal{F}}\mathcal{S}_{\mathcal{F}} &= 0 , \quad \text{if}\quad \mathcal{S}(\mathcal{F}) = 0 .
\end{aligned}
\tag{4.27}
$$

For $\mathcal{F}$ equal to the classical action $S$ we shall use the notation

$$
b := \mathcal{S}_S .
\tag{4.28}
$$

Since the classical action obeys the Slavnov–Taylor identity, the second of the nilpotency equations (4.27) implies the nilpotency of $b$:

$$b^2 = 0 \ . \tag{4.29}$$

From its definition one sees that the classical linearized Slavnov–Taylor operator $b$ acts in the following way on the fields[4]:

$$b\phi = s\phi \ , \qquad for \quad \phi = A, \ \psi, \ c, \ \bar{c}, \ B \ ,$$

$$b\hat{\rho}^\mu = \frac{\delta S}{\delta A_\mu} + \partial^\mu B = \frac{1}{g^2} D_\nu F^{\mu\nu} - \bar{\psi}\gamma^\mu T^a \psi \tau_a + i\{\hat{\rho}^\mu, c\} \ ,$$

$$bY = -\frac{\delta S}{\delta \bar{\psi}} = -i\slashed{D}\psi + ic^a T_a Y \ , \tag{4.30}$$

$$b\bar{Y} = \frac{\delta S}{\delta \psi} = i\bar{\psi}\slashed{D} - i\bar{Y} T_a c^a \ ,$$

$$b\sigma = \frac{\delta S}{\delta c} = \partial_\mu \hat{\rho}^\mu - i[A_\mu, \hat{\rho}^\mu] - i[\sigma, c] + i(\bar{Y} T^a \psi + \bar{\psi} T^a Y)\tau_a \ .$$

In order to prove the possibility of implementing the Slavnov–Taylor identity (4.18) to all orders we shall use the recursive procedure explained in Chap. 3. Assuming the problem to be solved till the order $n-1$ in $\hbar$, i.e., assuming the existence of a vertex functional obeying

$$\mathcal{S}(\Gamma) = O(\hbar^n) \ , \tag{4.31}$$

we want to look if a suitable choice of the counterterms at order $n$ makes it obey the next order identity

$$\mathcal{S}(\Gamma) = O(\hbar^{n+1}) \ . \tag{4.32}$$

Due to the quantum action principle in its general form (3.25), the action of the Slavnov–Taylor operator (4.13) on $\Gamma$ yields an integrated local insertion $\Delta \cdot \Gamma$ of ghost number 1 and of dimension 4 (cf. Table 4.1), of order $\hbar^n$ according to the recurrence assumption:

$$\mathcal{S}(\Gamma) = \hbar^n \Delta \cdot \Gamma = \hbar^n \Delta + O(\hbar^{n+1}) \ . \tag{4.33}$$

Since $\Gamma$ obeys the gauge condition (4.17) and the ghost equation (4.20) it follows from the application of (4.21) and (4.23) that the breaking $\Delta$ obeys the constraints:

$$\frac{\delta}{\delta B}\Delta = \mathcal{G}\Delta = 0 \ . \tag{4.34}$$

The first nilpotency property (4.27) together with

$$\mathcal{S}_\Gamma = b + O(\hbar) \ , \tag{4.35}$$

implies the *consistency condition*

$$b\Delta = 0 \tag{4.36}$$

on the integrated local field polynomial $\Delta(A, \psi, c, \bar{c}, \hat{\rho}, Y, \upsilon)$ which represents the first nonvanishing order of the breaking of the Slavnov–Taylor identity.

---

[4]Since we shall be working in a space of functionals obeying the ghost equation, it is convenient to write these transformation laws in terms of the shifted external field $\hat{\rho}^\mu$.

Let us suppose for a while that the general solution of the Consistency condition has the form

$$\Delta = b\tilde{\Delta} , \qquad (4.37)$$

where $\tilde{\Delta}$ is an integrated local field polynomial of dimension 4, (that this is a solution is the consequence of the nilpotency of $b$). The addition of the counterterm $-\hbar^n\tilde{\Delta}$ to the action alters the vertex functional as follows:

$$\Gamma \quad \rightarrow \quad \Gamma - \hbar^n\tilde{\Delta} + O(\hbar^{n+1}) .$$

Introducing this new vertex functional in the left–hand side of (4.33) yields:

$$S(\Gamma - \hbar^n\tilde{\Delta}) = S(\Gamma) - \hbar^n\Delta + O(\hbar^{n+1}) = O(\hbar^{n+1}) , \qquad (4.38)$$

which shows the validity of the Slavnov–Taylor identity (4.32) to the next order.

In order to complete the discussion we have to check that the addition of the counterterm $-\hbar^n\tilde{\Delta}$ preserves the gauge condition (4.17) and the ghost equation (4.20). It should therefore obey the constraints

$$\frac{\delta}{\delta B}\tilde{\Delta} = \frac{\delta}{\delta\bar{c}}\tilde{\Delta} = 0 . \qquad (4.39)$$

The derivative with respect to $\bar{c}$ in the second constraint is just the form of the ghost equation operator $\mathcal{G}$ after the change of variables defined by (4.24). In order to show that $\tilde{\Delta}$ indeed obeys these constraints, we first remark that the breaking $\Delta$ already obeys them (see (4.34)). On the other hand let us expand $\tilde{\Delta}$ according to the eigenvalues $n$ of the counting operator

$$\mathcal{N} := \mathrm{Tr}\int d^4x \left( B\frac{\delta}{\delta B} + \bar{c}\frac{\delta}{\delta\bar{c}} \right) = \left\{ b, \mathrm{Tr}\int d^4x\, \bar{c}\frac{\delta}{\delta B} \right\} , \qquad (4.40)$$

i.e., according to the degree of homogeneity in the fields $B$ and $\bar{c}$:

$$\tilde{\Delta} = \sum_{n\geq 0}\tilde{\Delta}_n , \quad \text{with} \quad \mathcal{N}\tilde{\Delta}_n = n\tilde{\Delta}_n .$$

The corresponding expansion for $\Delta$ reduces to the zeroth order term due to (4.34). Applying the operator $\mathcal{N}$ to both sides of (4.37) and using its commutativity with the BRS operator $b$ following from the second of eqs. (4.40) we obtain

$$0 = \sum_{n\geq 1}n\, b\tilde{\Delta}_n .$$

This implies $b\tilde{\Delta}_n = 0$ for any nonvanishing $n$, hence

$$\Delta = b\tilde{\Delta}_0 .$$

This is the desired result, since $\tilde{\Delta}_0$ being independent of $B$ and $\bar{c}$ obeys the constraints (4.39).

We have supposed up to now that the solution of the consistency condition (4.36) has the "trivial" form (4.37). But there may be solutions of the consistency condition which are not of this form. In this case the breaking of the Slavnov–Taylor identity cannot be absorbed as a counterterm and we have anomalies. This will be the subject of the next section.

*Remark.* The problem of solving the consistency condition (4.36) is a problem of *cohomology*. The coboundary operator is the nilpotent operator $b$. It acts in the space of local Poincaré invariant field polynomials, the "local functionals". The cohomology classes are defined by the equivalence relation:

$$\Delta' \sim \Delta'' \, , \tag{4.41}$$

if and only if    $\Delta' - \Delta'' = b\tilde{\Delta}$   , with $\tilde{\Delta}$ a local functional of dimension 4.

The class equal to zero is the set of trivial local functionals, i.e., those of the form (4.37). An anomaly is a nonvanishing cohomology class.

A detailed discussion of these aspects will be given in Chap. 5.

## 4.3 BRS Cohomology: The Gauge Anomaly

In this section we shall solve the consistency condition (4.36)

$$b\Delta(A, \psi, c, \bar{c}, \hat{\rho}, Y, \sigma) = 0 \, , \tag{4.42}$$

for $\Delta$ a local functional of dimension 4 and of ghost number 1. The coboundary operator $b$ is defined by (4.30).

We can make our task easier by noting that the classical action is invariant under the *rigid gauge transformations* ($\varepsilon^a = $ constants)

$$\delta_{\text{rig}}\phi(x) = i[\varepsilon^a \tau_a, \phi(x)] =: i\varepsilon^a \delta_a \phi(x) \, , \qquad \phi = A_\mu, c, \bar{c}, B, \hat{\rho}, \sigma \, ,$$
$$\delta_{\text{rig}}\psi(x) = i\varepsilon^a T_a \psi(x) =: i\varepsilon^a \delta_a \psi(x) \, , \tag{4.43}$$

and by relying on the proof, given in Subsect. 3.5.1, that such an invariance can be maintained to all orders. We may therefore assume the corresponding Ward identities to be fulfilled:

$$\mathcal{W}_a \Gamma =: \int d^4x \sum_\varphi \delta_a \varphi \frac{\delta}{\delta\varphi} \Gamma = 0 \, . \tag{4.44}$$

It will follow from the commutativity property

$$\mathcal{W}_a \mathcal{S}(\mathcal{F}) = \mathcal{S}_\mathcal{F} \mathcal{W}_a \mathcal{F} \, , \qquad \forall \mathcal{F} \, , \tag{4.45}$$

that the breaking $\Delta$ of the Slavnov–Taylor identity is rigid invariant:

$$\mathcal{W}_a \Delta = 0 \, . \tag{4.46}$$

As a consequence of the proposition 5.9 proved in Chap. 5 we can then restrict the cohomology problem to the space of rigid invariant local functionals.

*Remark.* Rigid invariance is a simplifying hypothesis, which is not necessary [16], and anyhow could not be imposed in more general cases, as for example in the case of a theory with spontaneously broken gauge invariance.

### 4.3.1 Elimination of the External Field and Matter Field Dependence from the Anomaly

Let us begin by showing that the dependence on the external fields $\hat{\rho}$, $Y$, and $\sigma$ is trivial, i.e., with the notation (4.41),

$$\Delta(A, \psi, c, \bar{c}, \hat{\rho}, Y, \sigma) \sim \Delta(A, \psi, c) . \tag{4.47}$$

It has been proved in quite generality [78, 79] that in a gauge theory such as the one considered in the present chapter the cohomology in the sector of ghost number 1 is independent from the external fields. This proof is valid for any semi–simple gauge group[5] and does not assume any restriction due to power–counting. But the latter allows a much simpler proof which we shall give here.

The most general dependence on $\sigma$, which has dimension 4 and ghost number $-2$ (see Table 4.1), is given by

$$\Delta = x \,\mathrm{Tr} \int d^4x \; \sigma c^3 \; + \; \text{terms independent of } \sigma .$$

The consistency condition (4.42) yields

$$
\begin{aligned}
0 = b\Delta \; &= x \,\mathrm{Tr} \int d^4x \; \left( \frac{\delta S}{\delta c} c^3 + i\sigma c^4 \right) \; + \; \text{terms independent of } \sigma \\
&= ix\,\mathrm{Tr} \int d^4x \; \left( [c, \sigma] c^3 + \sigma c^4 \right) \; + \; \text{terms independent of } \sigma \\
&= -ix\,\mathrm{Tr} \int d^4x \; \sigma c^4 \; + \; \text{terms independent of } \sigma ,
\end{aligned}
$$

hence $x = 0$, and $\Delta$ is independent of $\sigma$.

The most general $\bar{Y}$–dependence is given by

$$\Delta = y \int d^4x \; f^{abc} \bar{Y} T^a \psi c^b c^c \; + \; \text{terms independent of } \bar{Y}, \sigma . \tag{4.48}$$

One checks that (4.48) is the $b$–variation of

$$-2iy \int d^4x \; \bar{Y} T^a \psi c_a$$

up to $\bar{Y}$–independent terms. Hence $\Delta$ is independent of $\bar{Y}$ modulo the equivalence relation (4.41) (and of $Y$ as well).

At last the most general dependence in $\hat{\rho}$ is given by

$$\Delta \sim \mathrm{Tr} \int d^4x \; \hat{\rho}^\mu R_\mu(A, c) \; + \; \text{terms independent of } \hat{\rho}, Y, \sigma ,$$

$$R_\mu(A, c) = z \partial_\mu c c + t c \partial_\mu c + u A_\mu c^2 + v c A_\mu c + w c^2 A_\mu .$$

The consistency condition writes

$$
\begin{aligned}
0 = b\Delta \; &= \mathrm{Tr} \int d^4x \; \left( \frac{\delta S}{\delta A_\mu} - \hat{\rho}^\mu b R_\mu \right) \; + \; \text{terms independent of } \hat{\rho}, Y, \sigma , \\
&= \mathrm{Tr} \int d^4x \; \left( i\{\hat{\rho}^\mu, c\} R_\mu - \hat{\rho}^\mu b R_\mu \right) \; + \; \text{terms independent of } \hat{\rho}, Y, \sigma ,
\end{aligned}
$$

---

[5]If the gauge group contains Abelian factors, the cohomology may depend on the external fields [80, 81, 78, 79].

and yields after some calculations:

$$u = -iz , \quad v = 0 , \quad w = it .$$

It follows that $\Delta$ is the $b$–variation of

$$\mathrm{Tr} \int d^4x \; \hat{\rho}^\mu \left( -zA_\mu c + tcA_\mu \right) ,$$

up to terms independent of $\hat{\rho}$, $Y$ and $\sigma$. This ends the proof of (4.47).

Having thus eliminated the dependence on the external fields, let us write now the most general expression of dimension 4 depending on $\psi$:

$$\Delta \sim \int d^4x \; \bar{\psi}\gamma^\mu \left( xT^a\partial_\mu c_a + yT^aT^b c_a A_{\mu b} + zT^bT^a c_a A_{\mu b} + uT^a c_a \partial_\mu \right)\psi$$

$$+ \text{ terms dependent only on } A_\mu, c .$$

The application of the consistency condition yields the vanishing of all the coefficients except $x$:

$$y = z = u = 0 .$$

But the nonvanishing term is the $b$–variation of

$$x \int d^4x \; \bar{\psi}\gamma^\mu T^a\psi A_{\mu a}.$$

Thus

$$\Delta \sim \Delta(A, c) . \tag{4.49}$$

## 4.3.2 The Gauge Anomaly and the Descent Equations

The last step is to solve the consistency condition (where we have replaced $b$ by $s$ since both operators coincide on $A$ and $c$):

$$s\Delta(A, c) = 0 . \tag{4.50}$$

This is nothing else than the Wess–Zumino consistency condition [82], written in the BRS formalism. Writing[6]

$$\Delta = \int d^4x \; Q^1(x) , \tag{4.51}$$

we infer from the consistency condition that the BRS variation of the integrand must be a total derivative:

$$sQ^1 = \partial^\mu Q^2_\mu .$$

Nilpotency of $s$ implies now

$$\partial^\mu sQ^2_\mu = 0 ,$$

from which follows

$$sQ^2_\mu = \partial^\nu Q^3_{[\mu\nu]} ,$$

where $[\mu\nu \cdots]$ means antisymmetry in $\mu$, $\nu$, $\cdots$. The argument can be repeated and leads us to the *descent equations* [83, 84]

---

[6]The superscript indicates the ghost number.

$$sQ^1 = \partial^\mu Q^2_\mu$$
$$sQ^2_\mu = \partial^\nu Q^3_{[\mu\nu]}$$
$$sQ^3_{[\mu\nu]} = \partial^\rho Q^4_{[\mu\nu\rho]} \qquad (4.52)$$
$$sQ^4_{[\mu\nu\rho]} = \partial^\lambda Q^5_{[\mu\nu\rho\lambda]}$$
$$sQ^5_{[\mu\nu\rho\lambda]} = 0 \ .$$

Since $Q^5$ has dimension zero, and there is no field of negative dimension, the sequence of descent equations must stop here.

*Remark.* The antisymmetric tensors $Q^k$ are components of differential forms, and the right–hand sides of the descent equations are exterior derivatives of their duals. The validity of the argument above follows from the triviality of the cohomology of the exterior derivative in the space of local functionals (c.f. Prop. 5.10 in Chap. 5).

The solution of the descent equations (4.52) is worked out in Chap. 5, Subsect. 5.3.1. It is unique, up to terms which are either total derivatives or BRS variations, and is given by (5.43) in the language of differential forms (the $\Omega$'s, there, are the duals of the $Q$'s). In the notations of the present chapter the general solution for $Q^1$ reads

$$Q^1 = r\varepsilon_{\mu\nu\rho\sigma}\mathrm{Tr}\left(c\partial^\mu\left(\partial^\nu A^\rho A^\sigma + \frac{i}{2}A^\nu A^\rho A^\sigma\right)\right)$$
$$+\text{BRS variation} + \text{total derivative}$$
$$= r\varepsilon_{\mu\nu\rho\sigma}c_a\partial^\mu\left(d^{abc}\partial^\nu A^\rho_b A^\sigma_c + \frac{\mathcal{D}^{abcd}}{12}A^\nu_b A^\rho_c A^\sigma_d\right) \qquad (4.53)$$
$$+\text{BRS variation} + \text{total derivative} \ ,$$

with $r$ being some calculable coefficient depending on the parameters of the theory. $d_{abc}$ is the totally symmetric invariant tensor of rank 3 defined by

$$d_{(abc)} = \frac{1}{2}\mathrm{Tr}\left(\tau_a\{\tau_b, \tau_c\}\right) \ , \qquad (4.54)$$

and

$$\mathcal{D}_{abcd} = d^m{}_{ab}f_{ncd} + d^m{}_{ac}f_{ndb} + d^m{}_{ad}f_{nbc} \ .$$

Finally, integration on space–time yields the Adler–Bardeen nonabelian gauge anomaly, up to a BRS variation [85, 86]:

$$\Delta = r\mathcal{A} = r\varepsilon_{\mu\nu\rho\sigma}\mathrm{Tr}\int d^4x\, c\partial^\mu\left(\partial^\nu A^\rho A^\sigma + \frac{i}{2}A^\nu A^\rho A^\sigma\right)$$
$$= r\varepsilon_{\mu\nu\rho\sigma}\int d^4x\, c_a\partial^\mu\left(d^{abc}\partial^\nu A^\rho_b A^\sigma_c + \frac{\mathcal{D}^{abcd}}{12}A^\nu_b A^\rho_c A^\sigma_d\right) \ . \qquad (4.55)$$

*Remarks.*

1. The actual presence of an anomaly depends on the nonvanishing of its coefficient $r$. In the present case with one left–handed spinor field it is proportional to

$$\frac{1}{2}d_{abc}\mathrm{Tr}\left(T^a\{T^b,T^c\}\right)$$

in the one–loop approximation (triangle graph). The matter representation may be chosen such that this vanishes. In this case a nonrenormalization theorem [85, 87, 88, 89, 90] ensures its vanishing to all orders. We shall prove it in Sect. 6.3 of Chap. 6.

2. It is clear from the expression (4.55) that in theories not involving the tensor $\varepsilon_{\mu\nu\rho\sigma}$ (for example chromodynamics), or the tensor $d_{abc}$ (for example a SU(2) gauge theory) there is no room for an anomaly.

## 4.4  Stability

We have seen in the last section how noninvariant counterterms have in general to be included in the action, order by order, for the theory to remain BRS invariant. We have also seen that there may be an obstruction to this programme, namely the gauge anomaly. The anomaly will be assumed throughout this section to be absent.

We shall now show that the theory is stable under the effect of the radiative corrections, i.e., that the most general invariant counterterms correspond to a renormalization of the parameters already present in the classical approximation and to a mere redefinition of the field variables. This stability will be expressed intrinsically by the Callan–Symanzik equation, in the same way as in Chap. 3 for the scalar field theories.

### 4.4.1 Invariant Counterterms

The mechanism of the inclusion of the noninvariant counterterms as expressed by (4.37) and (4.38) shows that one is still free, at the order considered, to add any BRS invariant counterterm, i.e., an integrated local functional $L$ of ghost number 0 and of dimension not exceeding 4, obeying the condition

$$bL = 0 \ , \tag{4.56}$$

where $b$ is the classical linearized Slavnov–Taylor operator defined in (4.28). Solving the latter condition is again a problem of cohomology, similar to that of solving the consistency condition (4.36). Whereas the latter had to be solved in the space of integrated local functionals of ghost degree 1 and dimension 4, the former has to be solved in the sector of ghost number 0 and dimension 4. The calculation will be left as an exercise to the reader. The result is that the most general BRS invariant counterterm has the form

$$L = \mathrm{const} \cdot L_{\mathrm{cohom}} + b\hat{L} \ , \tag{4.57}$$

where

$$L_{\mathrm{cohom}} = \mathrm{Tr}\int d^4x \,\mathrm{Tr}\, F^{\mu\nu}F_{\mu\nu} \tag{4.58}$$

represents the unique element of the cohomology in the sector under consideration, and $\hat{L}$ is an arbitrary integrated local functional of ghost number $-1$ and dimension 4. The latter is the superposition of three field polynomials, shown in the following equations together with their $b$–variations[7]:

$$\hat{L}_1 = -\int d^4x \left(\bar{Y}\psi - \bar{\psi}Y\right) ,$$

$$b\hat{L}_1 = \mathcal{N}_\psi S := \left(N_\psi + N_{\bar{\psi}} - N_Y - N_{\bar{Y}}\right) S ,$$

$$\hat{L}_2 = \mathrm{Tr} \int d^4x \hat{\rho}^\mu A_\mu ,$$

$$b\hat{L}_2 = \mathcal{N}_A S := \mathrm{Tr} \left(N_A - N_\rho - N_B - N_{\bar{c}}\right) S + 2\alpha \partial_\alpha S ,$$

$$\hat{L}_3 = -\mathrm{Tr} \int d^4x \sigma c ,$$

$$b\hat{L}_3 = \mathcal{N}_c S := \mathrm{Tr} \left(N_c - N_\sigma\right) S .$$

(4.59)

The operators $N$ are the *counting operators*

$$N_\varphi := \int d^4x \varphi \frac{\delta}{\delta\varphi} , \quad \varphi = A, \psi, \bar{c}, B, \rho, Y, \sigma , \tag{4.60}$$

and $S$ is the classical action (4.14).

The four coefficients of the invariant counterterms may be fixed by normalization conditions similar to the ones introduced in Sects. 3.4 and 3.5 for the theories with rigid symmetry. For this we may take the four conditions

$$\frac{d}{dp^2}\tilde{\Gamma}^{\mathrm{T}}(p^2)\bigg|_{p^2=\mu^2} = -\frac{1}{g^2} ,$$

$$\frac{1}{4}\frac{\partial}{\partial p^\mu}\mathrm{Tr}\left(\gamma^\mu \tilde{\Gamma}_{\psi\bar{\psi}}(p)\right)\bigg|_{p^2=\mu^2} = 1 ,$$

$$\frac{d}{dp^2}\tilde{\Gamma}_{c\bar{c}}(p^2)\bigg|_{p^2=\mu^2} = 1 ,$$

$$\tilde{\Gamma}_{\sigma^a c^b c^c}(p_1, p_2, p_3)\bigg|_{p_i^2=\mu^2} = f_{abc} .$$

(4.61)

where the normalization parameter $\mu$ has mass dimension 1 and we have used the definition

$$\tilde{\Gamma}_{A^\mu A^\nu}(p) = \left(\delta_{\mu\nu} - \frac{p_\mu p_\nu}{p^2}\right)\tilde{\Gamma}^{\mathrm{T}}(p^2) + \frac{p_\mu p_\nu}{p^2}\tilde{\Gamma}^{\mathrm{L}}(p^2) .$$

## 4.4.2 Physical and Nonphysical Parameters

The four counterterms listed in (4.58) and (4.59) can be put in correspondence with the fields and the parameters of the classical action. Indeed the counterterm (4.58), which belongs to the cohomology, can be reabsorbed by a renormalization of the gauge coupling constant $g$, which is a *physical* parameter. On the other hand the three

---

[7]We have taken into account the gauge fixing condition (4.17) and the ghost equation (4.20), with the effect that there is no dependence on the Lagrange multiplier field $B$ and that the dependence on the antighost field and on the external field $\rho$ occurs only via the combination $\hat{\rho}$ (4.24).

counterterms (4.59), which are cohomologically trivial, can be reabsorbed by the three independent field redefinitions, compatible with BRS invariance,

$$A' = z_1 A \ , \rho' = -z_1\rho \ , B' = -z_1 B \ , \bar{c}' = -z_1\bar{c} \ ,$$

$$c' = z_2 c \ , \sigma' = -z_2\sigma \ , \tag{4.62}$$

$$\psi' = z_3\psi \ , Y' = -z_3 Y \ .$$

The renormalization of the gauge field $A$ is accompanied by a renormalization of the gauge parameter: $\alpha' = 2z_1\alpha$.

Let us recall that the field amplitude renormalizations (4.62) are not physical, as we have seen in Subsect. 2.2.6. The same holds for the gauge parameter as we shall explicitly check in Sect. 4.5. In the present context of gauge theories the nonphysical character of a renormalization is expressed by the cohomological triviality of the corresponding counterterm.

The set of physical parameters on the other hand is characterized by the cohomology in the sector of ghost number 0, which here has only one element and corresponds to the gauge coupling constant. If masses or selfcouplings of the matter fields (Yukawa couplings for instance) where present, they would also be related to nontrivial elements of the cohomology. (See [91] for the application to the Standard Model).

### 4.4.3 The Callan–Symanzik Equation

As we have already done for the models with rigid invariances of Chap. 3 we can study the response of the system to a change of scale. The scale being given in the present massless theory by the normalization parameter $\mu$, the generator of changes of scale is given by $\mu\partial/\partial_\mu$. We shall obtain the Callan–Symanzik equation by expanding the dimension 4 insertion defined by $\mu\partial_\mu\Gamma$ in a suitable basis of insertions. The symmetry properties of these insertions are given by those of $\mu\partial_\mu\Gamma$, which follow from the Slavnov–Taylor identity (4.18) (assumed to hold without anomaly), the gauge condition (4.17) and the ghost equation (4.20):

$$\mathcal{S}_\Gamma\left(\mu\frac{\partial\Gamma}{\partial\mu}\right) = 0 \ ,$$

$$\frac{\delta}{\delta B}\left(\mu\frac{\partial\Gamma}{\partial\mu}\right) = 0 \ , \tag{4.63}$$

$$\mathcal{G}\left(\mu\frac{\partial\Gamma}{\partial\mu}\right) = \left(\frac{\delta}{\delta\bar{c}} + \partial^\mu\frac{\delta}{\delta\rho^\mu}\right)\left(\mu\frac{\partial\Gamma}{\partial\mu}\right) = 0 \ ,$$

where $\mathcal{S}_\Gamma$ is the linearized Slavnov–Taylor operator defined by (4.22).

A basis of insertions of dimension 4 (and of ghost number 0) has been given in the classical approximation by (4.58) and (4.59). A quantum extension of this basis may be written in the form of a set of insertions generated by the action on the vertex functional of four independent *symmetric differential operators*. We define a symmetric operator as an operator $\nabla$ which is compatible with the Slavnov–Taylor identity, the gauge condition and the ghost equation, i.e., which fulfills the conditions

$$\nabla \mathcal{S}(\Gamma) - \mathcal{S}_\Gamma \nabla \Gamma = 0 \ ,$$

$$\frac{\delta}{\delta B} (\nabla \Gamma) = 0 \ , \tag{4.64}$$

$$\mathcal{G} (\nabla \Gamma) = 0 \ .$$

The four operators we are looking for are (see the definitions in (4.59), (4.60))

$$\frac{\partial}{\partial g} \ , \quad \mathcal{N}_\psi \ , \quad \mathcal{N}_A \ , \quad \mathcal{N}_c \ . \tag{4.65}$$

It is obvious that their action on $\Gamma$ reduces in the tree approximation to the four invariant classical insertions (4.58) and (4.59). Hence from Prop. 3.1 of Chap. 3 they constitute a basis for the quantum invariant insertions of dimension 4 and ghost number 0. The expansion of $\mu \partial_\mu \Gamma$ in this basis yields the Callan–Symanzik or renormalization group equation[8]

$$\mathcal{C}\Gamma := (\mu \partial_\mu + \beta_g \partial_g - \gamma_\psi \mathcal{N}_\psi - \gamma_A \mathcal{N}_A - \gamma_c \mathcal{N}_c) \, \Gamma = 0 \ . \tag{4.66}$$

The gauge $\beta$–function $\beta_g$ and the anomalous dimensions $\gamma_\varphi$ are computable from the normalization conditions (4.61) as power series in $\hbar$, in the same way as at the end of Sect. 3.4.2.

## 4.5 Gauge Independence of the Physical Operators

Assuming that the theory is anomaly–free, we shall show that a slightly extended Slavnov–Taylor identity implies the desired result that the physical quantities do not depend on the gauge fixing parameter $\alpha$ (cf. (4.8)).

Physical operators are defined classically as gauge invariant composite fields $Q^i(x)$ constructed from the basic fields $A_\mu$ and $\psi$. Their quantum counterparts are defined to be nontrivial BRS–invariant operators, i.e., $s$–cohomology classes, of zero ghost number. In the functional formalism this means that the insertions $Q^i(x)$ obey the Slavnov–Taylor identity

$$\mathcal{S}[Q_i(x) \cdot Z] = 0 \ , \tag{4.67}$$

with $\mathcal{S}$ given by (4.19), and that they are not of the trivial form

$$Q^i(x) \cdot Z = \mathcal{S}[\widetilde{Q}^i(x) \cdot Z] \ . \tag{4.68}$$

The construction of the "gauge invariant" operators $Q^i(x)$ is performed by introducing BRS–invariant external fields $q_i(x)$ coupled to them. One thus produces a Green functional $Z(J, q)$ obeying the Slavnov–Taylor identity

$$\mathcal{S}Z(J, q) = 0 \ , \tag{4.69}$$

with the same Slavnov–Taylor operator $\mathcal{S}$ as in (4.67).

Differentiation of the latter identity with respect to $q$ yields (4.67), since $\delta/\delta q$ commutes with $\mathcal{S}$.

---

[8]Both equations coincide since the present theory is massless. See Subsect. 3.4.3.

We shall now show that the Green functions of gauge invariant operators

$$\left\langle TQ^{i_1}(x_1) \cdots Q^{i_n}(x_n) \right\rangle ,$$  (4.70)

generated by the functional

$$Z_{\text{inv}}(q) = Z(J,q)|_{J=0} ,$$  (4.71)

do not depend on the gauge parameter $\alpha$:

$$\partial_\alpha Z_{\text{inv}}(q) = 0 .$$  (4.72)

The trick [92, 93] to be used is the extended BRS invariance, which consists in allowing the gauge parameter(s) to vary under BRS, keeping BRS nilpotency. In the present case this amounts to add, to the BRS transformation laws (4.9), the law

$$s\alpha = \chi , \quad s\chi = 0 ,$$  (4.73)

where $\chi$ is a Grassmann parameter, of ghost number one. Thus the Slavnov–Taylor identity (4.19) should be replaced by the extended Slavnov–Taylor identity

$$\mathcal{S}Z(J,q) = \mathcal{S}_{(\text{old})}Z(J,q) + \chi\frac{\partial}{\partial\alpha}Z(J,q) = 0 .$$  (4.74)

Differentiating this extended Slavnov–Taylor identity with respect to $\chi$ yields

$$\frac{\partial}{\partial\alpha}Z(J,q) - \mathcal{S}\frac{\partial}{\partial\chi}Z(J,q) = 0 .$$  (4.75)

Setting at the end $J = 0$ and $\chi = 0$ gives the gauge independence condition (4.72) we wanted.

The classical action which solves the extended Slavnov–Taylor identity in the tree–graph approximation is still given by the action (4.14), which however contains a supplementary, $\chi$–dependent, term

$$\frac{\chi}{2}\text{Tr}\int d^4x\ \bar{c}B .$$

Indeed, the gauge fixing part of the action being a BRS variation as the former one (4.15), acquires this $\chi$–dependence through the action of the BRS operator on $\alpha$.

Thus the problem of the gauge independence is reduced to the problem of implementing the extended Slavnov–Taylor identity to all orders. That this task is always achievable follows from the fact that the gauge parameters $\alpha$ and $\chi$ in view of (4.73) form a BRS doublet (see Def. 5.7 in Chap. 5). Prop. 5.8 applies and shows that the cohomology does not depend on $\alpha$ and $\chi$. Hence no other obstruction than the gauge anomaly – assumed to be absent – can occur.

On the other hand, if there appears the gauge anomaly (4.55), then the anomaly coefficient $r$ at its lowest nonvanishing order is independent from the gauge parameter $\alpha$ since $r\Delta$ belonging to the extended BRS cohomology cannot depend on $\alpha$.

For further use let us translate (4.74) and (4.75) in terms of the vertex functional $\Gamma$:

$$\mathcal{S}(\Gamma) := \mathcal{S}_{(\text{old})}(\Gamma) + \chi\frac{\partial\Gamma}{\partial\alpha} = 0 ,$$  (4.76)

with $\mathcal{S}_{(\text{old})}$ given by (4.13),

$$\frac{\partial}{\partial \alpha}\Gamma = \mathcal{S}_\Gamma \frac{\partial}{\partial \chi}\Gamma \;. \tag{4.77}$$

The linearized Slavnov–Taylor operator $\mathcal{S}_\Gamma$ is defined as in (4.22), but with the addition of the $\alpha$–$\chi$–term:

$$\mathcal{S}_\Gamma = \mathcal{S}_{\Gamma(\text{old})} + \chi \frac{\partial}{\partial \alpha} \;. \tag{4.78}$$

## 4.6 Gauge Invariance Together with Rigid Invariance

Up to the present point, the BRS formalism has been applied either for a gauge symmetry, in this chapter, or very briefly for a rigid symmetry in Sect. 3.1. We would like to show that this formalism can be particular powerful in the study of gauge theories constrained by a rigid invariance in addition to the local one. An example will be given in Chap. 7, where the rigid symmetry is a "hidden" one since acting in the ghost sector, manifesting itself as a supersymmetry whose generators commute nontrivially with the usual BRS operator. In the present section, we shall consider invariances acting on the physical fields. The BRS formalism will be used in order to implement the algebraic setup.

Concretely, we consider a Lie group $G$ and a subgroup $H \subset G$. The invariance under $H$ is local, whereas the invariance under the quotient $G/H$ is taken as rigid. Denoting by $X_a$ the generators of $H$ and by $Y_\alpha$ those of $G/H$, we postulate for $G$ the Lie algebra

$$[X_a, X_b] = if_{ab}{}^c X_c \;,$$

$$[Y_\alpha, X_b] = i\theta_{\alpha b}{}^c X_c \;, \tag{4.79}$$

$$[Y_\alpha, Y_\beta] = ig_{\alpha\beta}{}^\gamma Y_\gamma + ih_{\alpha\beta}{}^c X_c \;.$$

The absence of $Y$ terms in the first commutator follows from $H$ being a subgroup. The absence of such terms in the second commutator is necessary if one wants the $X$'s to generate gauge transformations and the $Y$'s rigid transformations. In other terms $H$ must be an invariant subgroup. A special case commonly met is the one where $G$ is a product $H \times H'$, i.e., where the rigid transformations form a group $H'$. It corresponds to the vanishing of the structure constants $\theta_{\alpha b}{}^c$ and $h_{\alpha\beta}{}^c$.

### 4.6.1 BRS Transformations

As usual to each local generator $X_a$ we associate a ghost field $c^a(x)$. Moreover, following the discussion of Sect. 3.1, to each rigid generator $Y_\alpha$ we associate a "global ghost" (constant ghost field) $\varepsilon^\alpha$. The Lie algebra structure (4.79) is expressed by the following BRS transformation laws of the ghosts:

$$sc^a(x) = -\tfrac{1}{2}f_{bc}{}^a c^b(x)c^c(x) - \theta_{\beta c}{}^a \varepsilon^\beta c^c(x) - \tfrac{1}{2}h_{\beta\gamma}{}^a \varepsilon^\beta \varepsilon^\gamma \;,$$

$$s\varepsilon^\alpha = -\tfrac{1}{2}g_{\beta\gamma}{}^\alpha \varepsilon^\beta \varepsilon^\gamma \;. \tag{4.80}$$

Denoting the gauge fields $A_\mu^a$ and matter fields $\psi$ by a common array $\phi^A$, we can write their BRS transformations – i.e., their infinitesimal transformations where the ghosts play the role of infinitesimal parameters – as:

$$s\phi^A(x) = t_a^A c^a(x) + ic^a(x)T_a{}^A{}_B\phi^B(x) + i\varepsilon^\alpha R_\alpha{}^A(x), \qquad (4.81)$$

where $t_a^A c^a = \partial_\mu c^a$ for $\phi^A = A_\mu^a$ and zero otherwise. Recall that the indices $a$, $b$, $\cdots$ refer to the generators of the gauge group, and hence to the gauge fields.

Whereas the gauge transformations are realized linearly by means of the matrices $(T_a)^A{}_B = T_a{}^A{}_B$, the rigid ones are realized nonlinearly, the $R_\alpha{}^A(x)$ being local functions of the fields $\phi^A(x)$. The algebraic structure of the problem is determined or recovered by requiring the nilpotency of the BRS operator $s$. The conditions $s^2 c^a = 0$ and $s^2\varepsilon^\alpha = 0$ yield the set of Jacobi identities – the generalization of (3.3)) –

$$\sum_{\text{cyclic perm. of } abc} f_{ab}{}^d f_{dc}{}^e = 0 \ ,$$

$$\theta_{ab}{}^d f_{dc}{}^e + \theta_{ac}{}^d f_{bd}{}^e - f_{bc}{}^d \theta_{ad}{}^e = 0 \ ,$$

$$\theta_{ab}{}^d \theta_{\gamma d}{}^e - \theta_{\gamma b}{}^d \theta_{ad}{}^e + g_{a\gamma}{}^\delta \theta_{\delta b}{}^e + h_{a\gamma}{}^d f_d{}^b e = 0 \ , \qquad (4.82)$$

$$\sum_{\text{cyclic perm. of } \alpha\beta\gamma} \left( h_{\alpha\beta}{}^d \theta_{\gamma d}{}^e + g_{\alpha\beta}{}^\delta h_{\gamma\delta}{}^e \right) = 0 \ ,$$

$$\sum_{\text{cyclic perm. of } \alpha\beta\gamma} g_{\alpha\beta}{}^\delta g_{\delta\gamma}{}^\varepsilon = 0 \ .$$

The first three identities show that the matrices

$$(X_a)_b{}^c = -if_{ab}{}^c \ , \qquad (Y_\alpha)_b{}^c = -i\theta_{\alpha b}{}^c$$

build a representation of the Lie algebra of $G$, with the Lie algebra of $H$ as representation space.

The condition $s^2\phi^A = 0$ gives

$$[T_a, T_b]^A{}_B = if_{ab}{}^c T_c{}^A{}_B \ , \qquad (4.83)$$

$$\frac{\delta R_\alpha{}^A(x)}{\delta\phi^C(y)} T_b{}^C{}_B - T_b{}^A{}_C \frac{\delta R_\alpha{}^C(x)}{\delta\phi^B(y)} = i\theta_{ab}{}^c T_c{}^A{}_B \ , \qquad (4.84)$$

$$\int dy \left( R_\alpha{}^C(y) \frac{\delta R_\beta{}^A(x)}{\delta\phi^C(y)} - R_\beta{}^C(y) \frac{\delta R_\alpha{}^A(x)}{\delta\phi^C(y)} \right) = ig_{\beta\alpha}{}^\gamma R_\gamma{}^A(x) + ih_{\beta\alpha}{}^c T_c{}^A{}_D\phi^D(x) \ , \qquad (4.85)$$

and

$$R_\alpha{}^b(x) = R_\alpha{}^b{}_c\phi^c(x) \ , \quad \text{with} \quad R_\alpha{}^b{}_c := i\theta_{\alpha c}{}^b \ . \qquad (4.86)$$

One recognizes in the three equations (4.83), (4.84) and (4.85) the realization – which is nonlinear for what concerns the generators belonging to $G/H$ – of the algebra (4.79). In addition, (4.86) says that the generators belonging to $G/H$ are constrained to act linearly on the gauge fields.

In the particular case of a completely linear realization of $G/H$, i.e., if

$$R_\alpha{}^B(x) = R_\alpha{}^B{}_C\phi^C(x) \ , \ \forall B \ , \qquad (4.87)$$

then (4.84) and (4.85) take the form

$$[R_\alpha, T_b]^A{}_B = i\theta_{ab}{}^c T_c{}^A{}_B \ ,$$
$$[R_\alpha, R_\beta]^A{}_B = ig_{\alpha\beta}{}^\gamma R_\gamma{}^A{}_B + ih_{\alpha\beta}{}^c T_c{}^A{}_B \ , \qquad (4.88)$$

which together with.(4.79) obviously define a matrix representation of (4.79).

## 4.6.2 Gauge Fixing

We have first to establish the BRS transformations of the Lagrange multiplier field $B_a$ and of the antighost field $\bar{c}_a$. We consider the case of the linear gauge fixing given by the condition (4.17). The rigid transformations acting linearly on the gauge field with the matrix $R_\alpha$ given by (4.86), they must also act linearly on $B$ and $\bar{c}$, with the transposed matrix $-R_\alpha^{\text{tr}}$, in order to assure the invariance of the gauge fixing action. This gives rise to the BRS transformations (in matrix notation)

$$s\bar{c} = B - i\varepsilon^\alpha \bar{c} R_\alpha \,,$$

$$sB = -i\varepsilon^\alpha B R_\alpha + \frac{i}{2}\varepsilon^\alpha \varepsilon^\beta h_{\alpha\beta}{}^c \bar{c} T_c \,.$$

(4.89)

The term quadratic in $\varepsilon$ is needed in order to ensure the nilpotency of $s$.

The complete gauge fixing action is constructed as a BRS variation according to (4.15):

$$
\begin{aligned}
S_{\text{gf}} &= s\operatorname{Tr}\int d^4x \left(\bar{c}\partial^\mu A_\mu + \frac{\alpha}{2}\bar{c}B\right)\\
&= \operatorname{Tr}\int d^4x \left(B\partial^\mu A_\mu + \frac{\alpha}{2}B^2 + \partial^\mu\bar{c}(\partial_\mu c + icTA_\mu) + \frac{i\alpha}{4}\varepsilon^\alpha\varepsilon^\beta h_{\alpha\beta}{}^c \bar{c}T_c\bar{c}\right)\,.
\end{aligned}
$$

(4.90)

## 4.6.3 Slavnov–Taylor Identity and Ward Identities

Denoting by $\rho_A$ and $\sigma_a$ the classical sources coupled to the BRS transforms of $\phi^A$ and $c^a$, respectively, we can write the Slavnov–Taylor identity expressing the invariance under the BRS transformations (4.80) and (4.89) as usually (no external source for the transformations of $B$ and $\bar{c}$ are needed since they are linear):

$$
\begin{aligned}
\mathcal{S}(\Gamma) := \int d^4x \left(\frac{\delta\Gamma}{\delta\rho_A}\frac{\delta\Gamma}{\delta\phi^A} + \frac{\delta\Gamma}{\delta\sigma_a}\frac{\delta\Gamma}{\delta c^a} + (B - i\varepsilon^\alpha\bar{c}R_\alpha)\frac{\delta\Gamma}{\delta\bar{c}}\right.\\
\left. + (-i\varepsilon^\alpha BR_\alpha + \frac{i}{2}\varepsilon^\alpha\varepsilon^\beta h_{\alpha\beta}{}^c\bar{c}T_c)\frac{\delta\Gamma}{\delta B}\right) - \tfrac{1}{2}g_{\beta\gamma}{}^\alpha\varepsilon^\beta\varepsilon^\gamma\frac{\partial\Gamma}{\partial\varepsilon^\alpha} = 0 \,.
\end{aligned}
$$

(4.91)

The problem of the renormalizability, i.e., the search of the anomalies and of the invariant counterterms is solved by computing the cohomology of the linearized Slavnov–Taylor operator $\mathcal{S}_S$ corresponding to the classical invariant action $S$. This encompasses in a same stroke both the gauge invariance and the rigid invariances.

The rest of the discussion depending on the particular model one wants to study, let us stop here the general discussion. However a more detailed investigation of the case where the symmetries belonging to $G/H$ act linearly turns out to be instructive. We thus assume for $R_\alpha{}^B$ the form given by (4.87). This implies that – in the Landau gauge $\alpha = 0$ – the $\varepsilon$–dependent terms of the action are linear in the quantum fields, which means that the equation

$$\frac{\partial\Gamma}{\partial\varepsilon^\alpha} = \int d^4x \left(-i\rho_A R_\alpha{}^A{}_B\phi^B + \sigma_a\left(iR_\alpha{}^a{}_b c^b - h_{\alpha\beta}{}^a\varepsilon^\beta\right)\right)\,,$$

(4.92)

in much the same way as the gauge condition (4.8) and the ghost equation (4.20), can be assumed to hold to all orders, fixing then the $\varepsilon$–dependence of $\Gamma$.

*Remark.* We have chosen the Landau gauge $\alpha = 0$. For a generic linear gauge, the right–hand side of (4.92) would contain a term quadratic in the quantum field $\bar{c}$, which would be subject to renormalization.

It follows that the vertex functional may be written as

$$\Gamma = \Gamma_0 + \int d^4x \left( -i\varepsilon^\alpha \rho_A R_\alpha{}^A{}_B \phi^B + i\varepsilon^\alpha \sigma_a R_\alpha{}^a{}_b c^b - \frac{1}{2}\sigma_a h_{\beta\gamma}{}^a \varepsilon^\beta \varepsilon^\gamma \right) , \qquad (4.93)$$

where $\Gamma_0$ does not depend on $\varepsilon$.

Introducing this expression into the Slavnov–Taylor identity (4.91) leads to terms of order 0, 1, 2 and 3 in $\varepsilon$: (the term of order 3 being proportional to the left–hand side of the fourth of the equations (4.82) vanishes).

$$\mathcal{S}(\Gamma) = \mathcal{S}_0(\Gamma_0) - i\varepsilon^\alpha \mathcal{W}_\alpha^{(G/H)} \Gamma_0 - \frac{1}{2}\varepsilon^\alpha \varepsilon^\beta h_{\alpha\beta}{}^a \left( \bar{\mathcal{G}}_a \Gamma_0 - \Delta^G_{a,\mathrm{cl}} \right) \qquad (4.94)$$

with

$$\mathcal{W}_\alpha^{(G/H)} = -\int d^4x \left( (R_\alpha \phi)\frac{\delta}{\delta\phi} - (\rho R_\alpha)\frac{\delta}{\delta\rho} + (R_\alpha c)\frac{\delta}{\delta c} - (\sigma R_\alpha)\frac{\delta}{\delta\sigma} \right.$$
$$\left. - (\bar{c}R_\alpha)\frac{\delta}{\delta\bar{c}} - (BR_\alpha)\frac{\delta}{\delta B} \right) , \qquad (4.95)$$

and

$$\bar{\mathcal{G}}_a = \int d^4x \left( \frac{\delta}{\delta c^a} - i\bar{c}_b T_a{}^b{}_c \frac{\delta}{\delta B_c} \right) , \qquad (4.96)$$
$$\Delta^G_{a,\mathrm{cl}} = -i\int d^4x \left( \rho_A T_a{}^A{}_B \phi^B + \sigma_b T_a{}^b{}_c c^c \right) .$$

Identifying to zero each term in the $\varepsilon$–expansion of (4.94) yields

*Order 0:* The usual Slavnov–Taylor identity for gauge invariance,

$$\mathcal{S}_0(\Gamma_0) = 0 , \qquad (4.97)$$

where $\mathcal{S}_0$ is the Slavnov–Taylor operator (4.91) taken at $\varepsilon = 0$.

*Order 1:* The Ward identities associated to the rigid invariance under the generators of $G/H$,

$$\mathcal{W}_\alpha^{(G/H)} \Gamma_0 = 0 , \qquad (4.98)$$

*Order 2:* The identity

$$\bar{\mathcal{G}}_a \Gamma_0 = \Delta^G_{a,\mathrm{cl}} , \qquad (4.99)$$

which has the form of a Ward identity with a linear breaking term in the right–hand side. This is the "antighost equation", which will be discussed in Chap. 6 (see Eq. (6.7)) in a more general context. It will be seen in fact that its validity does not require the present coset structure.

In the present case the antighost equation (4.99) follows directly from the generalized Slavnov–Taylor identity, but this comes out only if the structure constants $h_{\alpha\beta}{}^c$ are nonvanishing, i.e., only if the rigid transformations do not form a subgroup of the overall group $G$.

# 5. BRS Cohomology and Descent Equations

As we have seen in Chap. 4 the BRS cohomology in the space of the integrated local functionals of the fields translates, at the nonintegrated level, into the BRS cohomology modulo total space–time derivatives. This has led to a system of equations called the descent equations. In this chapter we shall systematize the study of the cohomology of a general BRS operator and of the corresponding descent equations. Beginning with some definitions and statements, we shall set up the general strategy for solving the descent equations, at the classical as well as at the renormalized level. The class of theories considered here is that of power counting renormalizable gauge theories, both of the Yang–Mills and of the topological type.

The presentation is made rather systematic, keeping in view the various applications which will be given later on.

The chapter is organized in the following way. The formalism is presented in Sects. 5.1 and 5.2. The classical approximation is solved in Sects. 5.3 and 5.4, and the renormalizability, i.e., the existence of a solution in perturbative quantum theory is proven in Sect. 5.5.

Unless otherwise stated, the gauge group is a general compact Lie group G.

## 5.1 The Descent Equations: Definitions and Generalities

### 5.1.1 Cohomology of $\delta$ Modulo $d$ and Cohomology of $\delta$

Let $\delta$ and $d$ be two nilpotent, mutually anticommuting, *coboundary* operators:

$$\delta^2 = d^2 = \{\delta, d\} = 0 . \tag{5.1}$$

They are assumed to act in a space $\mathcal{F}$ possessing a double grading. The elements of $\mathcal{F}$ are "forms" $\Omega_p^q$, their two degrees being given by the pair of indices $(q, p)$. The action of $\delta$, respectively of $d$, increases the degree $q$, respectively $p$, by one unit. In the applications given in this book, $\delta$ generally represents the BRS operator $s$ or its renormalized version $\mathcal{S}_\Gamma$, namely the linearized Slavnov–Taylor operator (4.78). In this case the upper index $q$ denotes the ghost number. On the other hand $d$ is the external derivative and $\Omega_p^q$ a differential form of degree $p$:

$$\Omega_p^q = \frac{1}{p!} \Omega_{\mu_1 \cdots \mu_p}^q(x) dx^{\mu_1} \cdots dx^{\mu_p} . \tag{5.2}$$

The set of forms can be divided into *even* and in *odd* forms, according to the parity of their *total degree* $q + p$. The coefficients $\Omega_{\mu_1 \cdots \mu_p}^q(x)$ are either local polynomials in the fields and their derivatives, or their renormalized version, i.e., quantum insertions. Since these coefficients anticommute with the differentials $dx^\mu$ for $q$ odd, a careful definition of the exterior derivative is needed: we choose the convention

$$d\Omega_p^q = \frac{1}{p!} dx^\nu \partial_\nu \Omega_{\mu_1 \cdots \mu_p}^q(x) dx^{\mu_1} \cdots dx^{\mu_p} . \tag{5.3}$$

Let us give some definitions.

**Definition 5.1** *A $\delta$-ladder $\Omega$ is a set of forms*

$$\Omega = \left\{ \Omega_p^{Q-p} \mid \Omega_p^{Q-p} \in \mathcal{F} , \ 0 \le p \le P , \ Q \ fixed \right\} , \tag{5.4}$$

*obeying the descent equations*

$$\delta\Omega_p^{Q-p} + d\Omega_{p-1}^{Q-p+1} = 0 , \quad 1 \le p \le P ,$$
$$\delta\Omega_0^Q = 0 . \tag{5.5}$$

*The total degree $Q$ is fixed. The degree $p$ takes all values between 0 and $P$, $P$ being lower or equal to the space–time dimension $D$. The $\delta$–ladder is said to be complete if $P = D$.*

The problem of solving the descent equations (5.5) is called a problem of *cohomology of $\delta$ modulo d*. The elements of the cohomology of $\delta$ modulo $d$ are the solutions of the descent equations which are not of the *trivial* form

$$\Omega_p^{Q-p} = \delta\hat{\Omega}_p^{Q-p-1} + d\hat{\Omega}_{p-1}^{Q-p} , \quad 1 \le p \le P ,$$
$$\Omega_0^Q = \delta\hat{\Omega}_0^{Q-1} , \quad \text{with} \quad \hat{\Omega}_{p-1}^{Q-p} \in \mathcal{F} , \tag{5.6}$$

where $\hat{\Omega}$ is some ladder of total degree $Q - 1$. More precisely:

**Definition 5.2** *Two $\delta$–ladders $\Omega$ and $\Omega'$ are equivalent if their difference is a trivial ladder of the type (5.6). The elements of the cohomology of $\delta$ modulo d are the corresponding equivalence classes.*

We shall also need in the following the concept of cohomology of $\delta$.

**Definition 5.3**

1. *A $\delta$–closed form $\Omega \in \mathcal{F}$, i.e., a form obeying the constraint*

$$\delta\Omega = 0 ,$$

   *is called a $\delta$–cocycle.*

2. *The cohomology of the coboundary operator $\delta$ in the space $\mathcal{F}$ of forms $\Omega$ is the set of equivalence classes for the equivalence relation*

$$``\ \Omega \sim \Omega' \quad if \ and \ only \ if \quad \Omega - \Omega' = \delta\hat{\Omega} \quad for \ some \quad \hat{\Omega} \in \mathcal{F}\ ",$$

   *where $\Omega$ and $\Omega'$ are two $\delta$–cocycles.*

## 5.1.2 The General Strategy for Solving the Descent Equations

The problem of solving the descent equations (5.5) in the classical approximation has been much studied in the literature [94, 95, 96, 97], where general theorems concerning the existence and the construction of the solutions are given. This concerns gauge theories in any space–time dimension [94, 95, 96], as well as supersymmetric theories [94, 96] and gravitation theories [96].

However, the extension to the quantum case is far less well known, so that the sketch of a general strategy is necessary.

The procedure begins from the bottom descent equation, i.e., by solving

$$\delta \Omega_0^Q = 0 \ . \tag{5.7}$$

This is a problem of cohomology of $\delta$ (see Def. 5.3). The solution of (5.7) thus writes

$$\Omega_0^Q = \mathcal{A}_0^Q + \delta \hat{\Omega}_0^{Q-1} \ ,$$

where $\mathcal{A}_0^Q$ is a representative of the cohomology of $\delta$. We next insert this result into the descent equation corresponding to the degree $p = 1$, getting

$$\delta \left( \Omega_1^{Q-1} - d\hat{\Omega}_0^{Q-1} \right) + d\mathcal{A}_0^Q = 0 \ .$$

In order to solve the latter equation we must assume the existence of a special solution $\tilde{\mathcal{A}}_1^{Q-1}$ of the descent equation at the 1-form level:

$$\delta \tilde{\mathcal{A}}_1^{Q-1} + d\mathcal{A}_0^Q = 0$$

Then

$$\delta \left( \Omega_1^{Q-1} - d\hat{\Omega}_0^{Q-1} - \tilde{\mathcal{A}}_1^{Q-1} \right) = 0 \ ,$$

which means again that we have to solve for the cohomology of $\delta$, finding:

$$\Omega_1^{Q-1} = \tilde{\mathcal{A}}_1^{Q-1} + \mathcal{A}_1^{Q-1} + d\hat{\Omega}_0^{Q-1} + \delta \hat{\Omega}_1^{Q-2} \ ,$$

where $\mathcal{A}_1^{Q-1}$ is a new representative of the cohomology of $\delta$. The procedure goes on iteratively. Let us suppose that, for some $p$ between 1 and $P - 1$, we have

$$\Omega_p^{Q-p} = \tilde{\mathcal{A}}_p^{Q-p} + \mathcal{A}_p^{Q-p} + d\hat{\Omega}_{p-1}^{Q-p} + \delta \hat{\Omega}_p^{Q-p-1} \ , \tag{5.8}$$

where $\mathcal{A}_p^{Q-p}$ is a representative of the cohomology of $\delta$ in the sector of forms of degrees $(Q - p, p)$, and $\tilde{\mathcal{A}}_p^{Q-p}$ depends on the cohomology of $\delta$ in the sectors corresponding to the degrees $\leq p - 1$. The next descent equation upwards then reads

$$\delta \Omega_{p+1}^{Q-p-1} + d \left( \tilde{\mathcal{A}}_p^{Q-p} + \mathcal{A}_p^{Q-p} \right) - \delta d\hat{\Omega}_p^{Q-p-1} = 0 \ .$$

In order to proceed we must assume now the existence of a special solution $\tilde{\mathcal{A}}_{p+1}^{Q-p-1} \in \mathcal{F}$ at the $(p + 1)$–form level:

$$\delta \tilde{\mathcal{A}}_{p+1}^{Q-p-1} + d \left( \tilde{\mathcal{A}}_p^{Q-p} + \mathcal{A}_p^{Q-p} \right) = 0 \ . \tag{5.9}$$

If this is the case, then (5.8) holds with $p$ replaced by $p + 1$:

$$\Omega_{p+1}^{Q-p-1} = \tilde{A}_{p+1}^{Q-p-1} + A_{p+1}^{Q-p-1} + d\hat{\Omega}_p^{Q-p-1} + \delta\hat{\Omega}_{p+1}^{Q-p-2} , \tag{5.10}$$

where $A_{p+1}^{Q-p-1}$ is a representative of the cohomology of $\delta$ in the sector of forms of degrees $(Q-p-1, p+1)$.

This concludes the general discussion of the solution of the descent equations.

*Remarks.*

1. Each level of the computation of the cohomology of $\delta$ modulo $d$, which involves the cohomology of $\delta$ at the same level, also depends on the cohomology of $\delta$ in the lower levels.

2. It may happen that the solvability condition (5.9) is not satisfied for some $p = p_0 \leq P$, for a generic parametrization of the cohomology of $\delta$ for $p \leq p_0$. This means that the descent equations can be solved only up to and including the degree $p_0$. One can nevertheless proceed further if one requires the condition (5.9), thus imposing constraints on the parameters of the cohomology for $p \leq p_0$. We shall meet such a case in Subsect. 5.3.1 (Eq. (5.36) and below)

We conclude this subsection by stating the following useful result, which completes the first remark, and which obviously follows from the discussion above:

**Proposition 5.4** *The cohomology of $\delta$ modulo $d$ vanishes if the cohomology of $\delta$ vanishes.*

## 5.2 Local Cohomologies

Let us call *local forms* the differential forms $\Omega$ whose coefficients $\Omega_{\mu_1\cdots\mu_p}(x)$ are *local functionals*, i.e., polynomials in the fields and their derivatives, all taken at the same point $x$. The integral of an $x$–dependent local form is also called a local functional.

**Definition 5.5** *The cohomologies of $\delta$, of the exterior derivative $d$ and of $\delta$ modulo $d$ in such a space of local forms are called local cohomologies.*

We shall for a while study exclusively the local cohomologies, going to the cohomology of quantum insertions in Sect. 5.5.

### 5.2.1 Filtration

The problem of solving the (local) cohomology of $\delta$ or that of $\delta$ modulo $d$ may be rather difficult in general. However, the following proposition is very helpful. It allows to replace, in a first step, the coboundary operator $\delta$ by a simpler one, $\delta_0$.

**Proposition 5.6** [94, 26, 52, 96] *Let $\mathcal{N}$ be a "filtration" operator mapping the space $\mathcal{F}$ of forms $\Omega$ into itself, the eigenvalues $n$ of $\mathcal{N}$ being the nonnegative integers. Let us* *suppose that the forms $\Omega_p^q$ and the coboundary operator $\delta$ can be expanded according to these eigenvalues:*

$$\Omega_p^q = \sum_n \Omega_{p(n)}^q \ , \qquad \mathcal{N}\Omega_{p(n)}^q = n\Omega_{p(n)}^q \ , \tag{5.11}$$

$$\delta = \sum_{n \geq 0} \delta_n \ , \qquad [\mathcal{N}, \delta_n] = n\delta_n \ . \tag{5.12}$$

*We assume that the filtration operator $\mathcal{N}$ commutes with the exterior derivative d. Then:*

*(1) $\delta_0$ is a coboundary operator: $\delta_0^2 = 0$ .*

*(2) The cohomology of $\delta$ is isomorphic to a subspace of the cohomology of $\delta_0$.*

*(3) The cohomology of $\delta$ modulo d is isomorphic to a subspace of the cohomology of $\delta_0$ modulo d.*

*Remarks.*

1. The choice of the most appropriate filtration operator $\mathcal{N}$ depends on the particular problem under consideration.

2. More general and powerful results exist [94], which allow in principle to compute exactly the cohomology $\delta$. However, the result above is quite sufficient in most cases, the exact answer being easily derived from the approximate one.

*Proof.* (We follow [96].) The proof of (1) is obvious. Let us prove (3) explicitly, the proof of (2) being quite analogous. We first show that it is always possible to choose in any cohomology class of $\delta$ modulo d a representative of the form $\Omega_p^q = \sum_{n \geq N} \Omega_{p(n)}^q$, in such a way that its leading term $\Omega_{p(N)}^q$ is a nonvanishing element of the $\delta_0$–cohomology modulo d. It is clear indeed that

$$\delta\Omega_p^q + d\Omega_{p-1}^{q+1} = 0 \quad \Rightarrow \quad \delta_0\Omega_{p(N)}^q + d\Omega_{p-1(N)}^{q+1} = 0 \ .$$

Furthermore, if $\Omega_{p(N)}^q$ is ($\delta_0$ modulo d)–trivial, i.e., of the form

$$\Omega_{p(N)}^q = \delta_0\hat{\Omega}_{p(N)}^{q-1} + d\hat{\Omega}_{p-1(N)}^q \ ,$$

one can write

$$\Omega_p^q = \delta\hat{\Omega}_{p(N)}^{q-1} + d\hat{\Omega}_{p-1(N)}^q \quad + \text{ terms of order } n \geq N + 1 \ ,$$

which means that $\Omega_p^q$ is equivalent, in the sense of Definition 5.2, to a form with a leading term of order $N + 1$. The procedure can be continued till the lowest order term becomes nontrivial.

We must still show that the map from the cohomology of $\delta$ modulo d to the cohomology of $\delta_0$ modulo d thus introduced can be defined as an injective map, i.e., a map such that if $\Omega_p^q$ and $\Omega'^q_p$ are representatives of two distinct cohomology classes of $\delta$ modulo $d$, then $\Omega^q_{p(N)}$ and $\Omega'^q_{p(N')}$ belong to two distinct cohomology classes of $\delta_0$ modulo $d$. This is obvious for $N \neq N'$. For $N = N'$, suppose the converse, i.e.,

$$\Omega^q_{p(N)} = \Omega'^q_{p(N)} + \delta_0 \hat{\Omega}^{q-1}_{p(N)} + d\hat{\Omega}^q_{p-1(N)} \ .$$

Then (see Definition 5.2)

$$\Omega^q_p - \Omega'^q_p \sim \sum_{n \geq N+1} \Omega''^q_{p(n)} \ ,$$

which allows us to take, for example, $\Omega^q_p$ and $\Omega''^q_p$ as independent representatives, instead of $\Omega^q_p$ and $\Omega'^q_p$. Repeating the operation if needed, it is clear then that one arrives to a pair $\Omega^q_p$, $\Omega''^q_p$, whose leading terms $\Omega^q_{p(N)}$ and $\Omega''^q_{p(N'')}$ are not ($\delta_0$ modulo $d$)–equivalent, since $N''$ is not lower than $N+1$. This ends the proof of Prop. 5.6. ∎

### 5.2.2 BRS Doublets

The problem of computing a cohomology can be greatly simplified thanks to a very simple proposition which leads to the elimination from the game of all the fields appearing as BRS doublets.

**Definition 5.7** *A pair of fields* $(u, v)$ *is called a BRS doublet, or $\delta$–doublet, if*

$$\delta u = v \ , \qquad \delta v = 0 \ . \tag{5.13}$$

**Proposition 5.8** [94, 26, 96] *If a set of pairs of fields* $(u_i, v_i)$ *appear as $\delta$–doublets, then the cohomology of $\delta$ does not depend on these fields.*

*Proof.* (We follow [96].) Let us introduce the two operators

$$P = \sum_i \int d^4 x \left( u_i \frac{\delta}{\delta u_i} + v_i \frac{\delta}{\delta v_i} \right) \qquad \text{and} \qquad Q = \sum_i \int d^4 x \ u_i \frac{\delta}{\delta v_i} \ . \tag{5.14}$$

They obey the relations

$$P = \{\delta, Q\} \ , \qquad [\delta, P] = 0 \ . \tag{5.15}$$

In order to solve the $\delta$–cohomology

$$\delta \Delta = 0 \ ,$$

let us expand $\Delta$ in eigenvectors of $P$:

$$\Delta = \sum_{n \geq 0} \Delta_n \ , \qquad P\Delta_n = n\Delta_n \ . \tag{5.16}$$

It is clear that $\delta \Delta_n = 0$, $\forall n$. From (5.16) and (5.15) then follows

$$\Delta = \Delta_0 + \sum_{n>0} \frac{1}{n} P \Delta_n$$
$$= \Delta_0 + \sum_{n>0} \frac{1}{n} \delta\, Q \Delta_n = \Delta_0 + \delta(\cdots) \, .$$

This shows that the nontrivial part of $\Delta$ is contained in $\Delta_0$. But the latter is independent of the fields $u_i$, $v_i$ since $P\Delta_0 = 0$. This ends the proof.    ∎

*Remark.* $u_i$ and $v_i$ may also be constant parameters. An example of such a BRS doublet is provided, in Sect. 4.5, by the pair formed of the gauge parameter $\alpha$ and of its associated Grassmann parameter $\chi$, with the BRS transformation law (4.73). Gauge parameter independence was shown there to be a consequence of the cohomological triviality of the BRS–doublet dependent cocycles.

### 5.2.3 Restriction to a Subspace Constrained by a Rigid Symmetry

It very often happens that the space of (local) functionals $\Delta$ in which one has to solve the cohomology equation

$$\delta\Delta = 0 \tag{5.17}$$

is restricted by some condition $C$ on $\Delta$. If one decides to call trivial any coboundary $\delta\hat{\Delta}$ with $\hat{\Delta}$ not constrained by $C$, the constraint on $\Delta$ certainly lowers the dimension of the cohomology, i.e., the number of independent nontrivial solutions. But it is natural to impose the same condition $C$ on $\hat{\Delta}$ as on $\Delta$, in particular if it is a symmetry of the theory. In general this increases the dimension of the cohomology. Such a constrained cohomology is called equivariant cohomology [98].

We shall however consider here an example where the constraint on $\hat{\Delta}$ will be shown to follow from that on $\Delta$, so that the dimension of the cohomology is not changed. This example, both of interest for the applications and easy to solve (c.f. Appendix A of [99]), is the one where the constraint $C$ is the invariance under a group of rigid transformations. This may be the rigid symmetry obtained by restricting the gauge transformations to space–time independent ones (see (4.43)). This may also be another symmetry group, which we shall then assume to be a compact Lie group commuting with the gauge transformations, and more generally with the BRS operator $\delta$.

The problem may then be formulated as follows: compute the cohomology of $\delta$ for the local functionals constrained by

$$\delta_a^{\text{rig}} \Delta = 0 \, , \tag{5.18}$$

where the index $a$ labels the generators of the rigid symmetry group, in the case a trivial element is defined by

$$\Delta = \delta\hat{\Delta} \, , \tag{5.19}$$

with

$$\delta_a^{\text{rig}} \hat{\Delta} = 0 \, . \tag{5.20}$$

The solution is given by the following proposition:

**Proposition 5.9** *The cohomology of $\delta$ in the space of local functionals obeying the constraint (5.18) is given by selecting, among the elements of the unconstrained cohomology, those which obey the constraint – i.e., by selecting the equivalence classes which contain at least one element obeying the constraint.*

*Proof.* The proof consists in showing that if $\Delta$ in the equation

$$\Delta = \delta\hat{\Delta} , \qquad (5.21)$$

obeys (5.18), then $\hat{\Delta}$ can be chosen to obey (5.20).

The local functional $\hat{\Delta}$ can be decomposed as

$$\hat{\Delta} = \hat{\Delta}^{\natural} + \hat{\Delta}^{\flat} , \qquad (5.22)$$

where $\hat{\Delta}^{\natural}$ is rigid invariant,

$$\delta_a^{\mathrm{rig}}\hat{\Delta}^{\natural} = 0 ,$$

i.e., it belongs to the trivial representation of the group, and $\hat{\Delta}^{\flat}$ belongs to a representation which can be decomposed in a direct sum of nontrivial irreducible representations. Let us expand the latter according to the eigenvalues of the quadratic Casimir operator $C = \sum_a (\delta_a^{\mathrm{rig}})^2$:

$$\hat{\Delta}^{\flat} = \sum_i \hat{\Delta}_i , \quad \text{with} \ \ C\hat{\Delta}_i = c_i\hat{\Delta}_i , c_i \neq 0 . \qquad (5.23)$$

The corresponding decomposition of $\Delta$ contains only the invariant part $\Delta^{\natural}$ by assumption. The application of the Casimir operator to (5.21) yields, due to the commutativity of $\delta_a^{\mathrm{rig}}$ with $\delta$, the condition

$$0 = \sum_i c_i\delta\hat{\Delta}_i ,$$

which is equivalent to

$$\delta\hat{\Delta}_i = 0 , \ \forall i , \qquad (5.24)$$

since the eigenvalues $c_i$ are all nonvanishing and the $\hat{\Delta}_i$ are linearly independent. This allows to rewrite (5.21) as

$$\Delta = \delta\hat{\Delta}^{\natural} ,$$

which is the announced result. ∎

### 5.2.4 Poincaré Lemma for $d$

The study of the local cohomology of $d$, in contrast with the usual de Rham cohomology [100], is a purely algebraic problem, completely solved by the so–called "Algebraic Poincaré Lemma", which we state here without proof (see [96], and also [101, 94]):

**Proposition 5.10** *The local cohomology of the exterior derivative $d$ consists only of the constant 0–forms and of the $D$–forms*

$$\mathcal{L}(x)d^D x , \qquad (5.25)$$

*where $D$ is the space–time dimension and $\mathcal{L}(x)$ is an arbitrary local functional of the component fields with a nonvanishing Euler–Lagrange derivative.*

The problem of the cohomologies of $\delta$ and of $\delta$ modulo $d$ will be the object of the rest of the present chapter.

## 5.3 General Solution of the Classical Descent Equations in Yang–Mills Theory

The aim of the present and of the following sections is the computation of the cohomology of $\delta$ modulo $d$ in various situations of interest for gauge theories. The ladders we shall consider are made of local forms (c.f. Subsect. 5.2.4)

$$\Omega_p^{Q-p} = \frac{1}{p!}\Omega_{\mu_1\cdots\mu_p}^{Q-p}(x)dx^{\mu_1}\cdots dx^{\mu_p} , \quad 0 \leq p \leq P ,$$
$$Q, P \text{ fixed} , \quad P \leq D = \text{space–time dimension} ,$$

(5.26)

of the Yang–Mills field $A_\mu$ and of the Faddeev–Popov ghost $c$ of the gauge theories discussed in Chap. 4. The upper index denotes now the ghost number.

Moreover, as in Chap. 4, we shall assume that all objects are invariant under the rigid symmetry (4.43).

The problem for which we want to sketch a solution in the present section is that of the descent equations (5.5), where we take for the coboundary operator $\delta$ the BRS operator $s$ defined by (4.9). The descent equations read now

$$s\Omega_p^{Q-p} + d\Omega_{p-1}^{Q-p+1} = 0 , \quad 1 \leq p \leq P ,$$
$$s\Omega_0^Q = 0 ,$$

(5.27)

expressing the problem of the local cohomology of $s$ modulo $d$.

As it was claimed in Subsect. 4.3.2, the descent equation at degree $p - 1$ follows from the one at degree $p$ as a consequence of the algebraic Poincaré lemma (Prop. 5.10) for the exterior derivative $d$ in the space of local functionals. The reversed path, i.e., going upwards from the degree $p$ to the degree $p + 1$, will depend on the local cohomology of $s$, as we shall see.

### 5.3.1 The Descent Equations of the Gauge Anomaly

General results on the cohomologies of $s$ and of $s$ modulo $d$ will be stated in the following subsection. We shall now solve explicitly the case $Q = 5$, $P = D = 4$ for a simple Lie group:

$$s\Omega_p^{5-p} + d\Omega_{p-1}^{6-p} = 0 , \quad 1 \leq p \leq 4 ,$$
$$s\Omega_0^5 = 0 .$$

(5.28)

This corresponds to the solution of the cohomology giving rise to the anomaly of four–dimensional Yang–Mills theories, as announced in Chap. 4, where it was shown that the integrand of the anomaly (4.51) obeys the descent equations (4.52), once the external fields and the matter fields were eliminated. The equations (5.28) are their translation in the language of differential forms.

Defining the Yang–Mills connection form

$$A = A_\mu(x)dx^\mu , \tag{5.29}$$

we can rewrite the action of $s$ on the 1–form $A$ and on the 0–form $c$ as[1]:

$$sA = -Dc := -(dc - i[c, A]) , \quad sc = ic^2 . \tag{5.30}$$

Since four–dimensional Yang–Mills theory is powercounting renormalizable, with the consequence that the anomaly has dimension 4 ($A_\mu$ and $c$ have dimensions 1 and 0, respectively) the possible solutions of the descent equations will be constrained by the corresponding requirement that the dimension of the $p$–form $\Omega_p^{5-p}$ be equal to $p$.

We shall use Prop. 5.6, with the filtration operator

$$\mathcal{N} = \int d^4x \left( A(x)\frac{\delta}{\delta A(x)} + c(x)\frac{\delta}{\delta c(x)} \right) . \tag{5.31}$$

(This operator counts the total number of fields in a given monomial). The BRS operator decomposes accordingly as

$$s = s_0 + s_1 , \tag{5.32}$$

with

$$s_0 A = -dc , \quad s_0 c = 0 ; \quad s_0^2 = 0 . \tag{5.33}$$

Thus we begin by solving the descent equations (5.28) with $s$ replaced by $s_0$:

$$s_0 \Omega_p^{5-p} + d\Omega_{p-1}^{6-p} = 0 , \quad 1 \le p \le 4 ,$$
$$s_0 \Omega_0^5 = 0 , \tag{5.34}$$

Let us first find the local cohomology of $s_0$, in the space of the fields $A_\mu$, $c$ and of their derivatives. It is given by the following proposition:

**Proposition 5.11** *The local cohomology of $s_0$ for the local functionals of $A_\mu$ and $c$ consists of the field polynomials generated by the skew–symmetric tensor $\partial_{[\mu}A_{\nu]}$ and its derivatives, and by $c$ without derivative.*

*Proof.* [96] The local functionals $F(x)$ are polynomials in the fields $c$ and $A_\mu$ and in their derivatives. For the latter we shall use the notation

$$c_{,\mu_1\cdots\mu_n} := \partial_{\mu_1}\cdots\partial_{\mu_n}c ,$$
$$A_{[\mu,\mu_1]\mu_2\cdots\mu_n} := \tfrac{1}{2}\partial_{\mu_2}\cdots\partial_{\mu_n}(\partial_{\mu_1}A_\mu - \partial_\mu A_{\mu_1}) , \tag{5.35}$$
$$A_{(\mu,\mu_1)\mu_2\cdots\mu_n} := \tfrac{1}{2}\partial_{\mu_2}\cdots\partial_{\mu_n}(\partial_{\mu_1}A_\mu + \partial_\mu A_{\mu_1}) .$$

Since no partial integration is possible, the field derivatives have to be considered as independent fields. The $s_0$–transformations read

---

[1]The bracket $[X,Y]$ denotes the *graded commutator* of $X$ and $Y$. It is an anticommutator if both $X$ and $Y$ are odd, and a commutator otherwise.

$$s_0 A_{(\mu,\mu_1)\mu_2\cdots\mu_n} = c_{,\mu\mu_1\cdots\mu_n} \ , \quad n \geq 0 \ ,$$

$$s_0 c_{,\mu\mu_1\cdots\mu_n} = 0 \ , \qquad\qquad n \geq 0 \ ,$$

$$s_0 A_{[\mu,\mu_1]\mu_2\cdots\mu_n} = 0 \ , \qquad n \geq 1 \ ,$$

$$s_0 c = 0 \ .$$

One sees that the pairs of symmetric derivatives $\left(A_{(\mu,\mu_1)\mu_2\cdots\mu_n}, c_{,\mu\mu_1\cdots\mu_n}\right)$ for $n \geq 0$ are doublets of $s_0$. It follows from Prop. 5.8 that the cohomology of $s_0$ is independent of them, which is the statement of Prop. 5.11. ∎

With the help of this proposition we can solve the descent equations (5.34) following the lines sketched in Subsect. 5.1.1. Beginning with the equation for the 0–form $\Omega_0^5$, taking into account its ghost number and rigid invariance we are left with

$$\Omega_0^5 = \frac{x}{5}\operatorname{Tr} c^5 \ , \tag{5.36}$$

where $x$ is an arbitrary coefficient. The equation for the 1–form then writes

$$0 = s_0 \Omega_1^4 + x\operatorname{Tr} dcc^4 = s_0 \left( \Omega_1^4 - x\operatorname{Tr}\left(Ac^4\right)\right) \ . \tag{5.37}$$

The local cohomology of $s_0$ in this sector being trivial the latter equation solves to

$$\Omega_1^4 = s_0 \hat{\Omega}_1^3 + x\operatorname{Tr}\left(Ac^4\right) \ .$$

Taking into account the fact that[2]

$$d\operatorname{Tr}\left(Ac^4\right) = -s_0 \operatorname{Tr}\left(A^2 c^3 + AcAc^2\right) + \operatorname{Tr}\left(dAc^4\right) \ ,$$

the descent equation for the 2–form reads

$$0 = s_0 \left( \Omega_2^3 - d\hat{\Omega}_1^3 - x\operatorname{Tr}\left(A^2 c^3 + AcAc^2\right)\right) + x\operatorname{Tr}\left(dAc^4\right) \ . \tag{5.38}$$

Since the last term belongs to the cohomology of $s_0$, whereas the first one is trivial, both must vanish separately. Hence

$$x = 0 \ ,$$

and (5.38) solves to

$$\Omega_2^3 = d\hat{\Omega}_1^3 + s_0 \hat{\Omega}_2^2 + y\operatorname{Tr}\left(dAc^3\right) + z\operatorname{Tr}\left(*dAc^3\right) \ .$$

Here $y$ and $z$ are the arbitrary coefficients of the two cohomology elements in this sector[3]. Using

---

[2] The symbol $\wedge$ for the products of forms will be omitted.

[3] The symbol $*$ denotes the Hodge duality, defined for any $p$-form $\omega$ in $D$–dimensional space by

$$*\omega = \frac{1}{(D-p)!} \left(*\omega\right)_{\beta_1\beta_2\cdots\beta_{D-p}} dx^{\beta_1} dx^{\beta_2} \cdots dx^{\beta_{D-p}} \ , \tag{5.39}$$

where

$$\left(*\omega\right)_{\beta_1\beta_2\cdots\beta_{D-p}} = \frac{1}{p!} \varepsilon_{\beta_1\beta_2\cdots\beta_{D-p}\alpha_1\cdots\alpha_p} \omega^{\alpha_1\cdots\alpha_p} \ .$$

$$d(dAc^3) = -s_0 \mathrm{Tr}\, (dA(Ac^2 + cAc + c^2A)) \,,$$

$$d(*dAc^3) = -s_0 \mathrm{Tr}\, (*dA(Ac^2 + cAc + c^2A)) + d * dAc^3 \,,$$

we find that the descent equation for the 3–form reads

$$0 = s_0 \left(\Omega_3^2 - d\hat{\Omega}_2^2 - y\mathrm{Tr}\,(dA(Ac^2 + cAc + c^2A)) - z\mathrm{Tr}\,(*dA(Ac^2 + cAc + c^2A))\right)$$
$$+ z\mathrm{Tr}\,(d * dAc^3) \,.$$

$$(5.40)$$

Here again the right–hand side is the sum of a trivial term and of an element of the cohomology. Hence

$$z = 0 \,,$$

and we can solve (5.40):

$$\Omega_3^2 = d\hat{\Omega}_2^2 + s_0\hat{\Omega}_3^1 + y\mathrm{Tr}\,\left(dA(Ac^2 + cAc + c^2A)\right) + t\mathrm{Tr}\,(d * dAc^2) \,,$$

where $t$ is the coefficient of the single element of the cohomology in the present sector.

By similar computations we find for the descent equation of the 4–form:

$$0 = s_0 \left(\Omega_4^1 - d\hat{\Omega}_3^1 - t\mathrm{Tr}\,(d * dA(Ac + cA)) - y\mathrm{Tr}\,(dA(A^2c + AcA + cA^2))\right)$$
$$+ 2y\mathrm{Tr}\,(dAdAc^2) \,,$$

$$(5.41)$$

from which we conclude that

$$y = 0 \,,$$

since the last term belongs to the cohomology of $s_0$. Thus

$$\Omega_4^1 = d\hat{\Omega}_3^1 + s_0\hat{\Omega}_4^0 + u\mathrm{Tr}\,(dA * dAc) + r\mathrm{Tr}\,(dAdAc) \,,$$

$$(5.42)$$

where $u$ and $r$ are the coefficients of the two elements of the cohomology of $s_0$ in this sector. Notice that in the derivation of (5.42), use has been made of the fact that the term $\mathrm{Tr}\,(d * dA(Ac + cA))$ in (5.41) is trivial, i.e.,

$$\mathrm{Tr}\,(d * dA(Ac + cA)) = d\mathrm{Tr}\,(*dA(Ac + cA)) + s_0\mathrm{Tr}\,(*dAA^2) \,.$$

We have thus found that the local cohomology of $s_0$ modulo $d$ consists of two elements. Representatives of them are given by the two last terms of (5.42). It follows from Prop. 5.6 that the cohomology of $s$ modulo $d$ has at most two elements. But it is easy to check that the term of coefficient $u$ cannot be extended to a ladder solving the descent equations (5.28), whereas the extension of the term with coefficient $r$ exists. Thus the cohomology of $s$ modulo $d$ defined by the descent equations (5.28) has a unique element, a representative of which is given by the following $s$–ladder [84]:

$$\Omega_4^1 = \frac{1}{2}\mathrm{Tr}\,((AdA + dAA - iA^3)dc) \,,$$
$$\Omega_3^2 = \mathrm{Tr}\,(Adcdc) \,,$$
$$\Omega_2^3 = \mathrm{Tr}\,(cdcdc) \,,$$
$$\Omega_1^4 = \frac{i}{2}\mathrm{Tr}\,(c^3dc) \,,$$
$$\Omega_0^5 = -\frac{1}{10}\mathrm{Tr}\,c^5 \,.$$

$$(5.43)$$

Another interesting example [102] is the case $Q = 3$ in 4 dimensions:

$$\Omega_3^0 = \text{Tr} \left( AdA - i\frac{2}{3}A^3 \right) \ ,$$

$$\Omega_2^1 = \text{Tr} \left( Adc \right) \ ,$$

$$\Omega_1^2 = \text{Tr} \left( cdc \right) \ , \tag{5.44}$$

$$\Omega_0^3 = \frac{i}{3}\text{Tr} \, c^3 \ ,$$

which we leave as an exercise to the reader. This ladder plays a fundamental role in the discussion of the topological Chern–Simons theory (c.f. Chap. 7), and also in the proof of the nonrenormalization of the U(1) anomaly in nonabelian gauge theories [102].

*Remark.* If the gauge group is the Abelian group U(1), then the cohomology of $s$ modulo $d$ consists of two $s$–cocycles: the terms of coefficients $u$ and $r$, respectively, in (5.42). This is the case e.g. for the theory of a fermion field coupled to a U(1) axial vector field [69]. However, here, parity invariance rules out the appearance of the integral of the first cocycle as an anomaly, leaving the axial anomaly $\int dAdAc$. Similarly, in quantum electrodynamics, parity forbids the axial anomaly, whereas invariance under charge conjugation forbids the former one [69].

### 5.3.2 General Results for Yang–Mills Theories in the Classical Approximation

We state here without proof some results for the Yang–Mills theories in $D$–dimensional space–time. Such theories being power–counting nonrenormalizable if $D > 4$, we consider then the gauge field $A$ and its ghost $c$ as external fields, the components $A_\mu^a$ being coupled to currents of the quantum matter fields. This amounts to study the algebra of the current operators [86, 103, 17, 84]. The gauge anomaly will then correspond to a breakdown of current conservation and to the appearance of Schwinger terms in the current commutators.

### 5.3.3 Local Cohomology of $s$

We consider here again the case of local functionals depending on the gauge field $A_\mu^a$ and its ghost $c$, but now in $D$–dimensional space–time. The gauge group is supposed to be a compact Lie group G, and the BRS transformations are given by (5.30).

**Proposition 5.12** [94, 95, 96] *The local cohomology of $s$ consists of the vector space generated by linear superpositions of the factorized elements*

$$\mathcal{Q}_{\text{inv}}(c)\,\mathcal{R}_{\text{inv}}(F_{\mu\nu}, D_{\rho_1} \cdots D_{\rho_n} F_{\mu\nu}) \ , \tag{5.45}$$

*where the factors $\mathcal{Q}_{\text{inv}}$ and $\mathcal{R}_{\text{inv}}$ are G–invariant polynomials in $c$ (without derivatives), and in the Yang–Mills strength and its covariant derivatives, respectively.*

*Remark.* Although stated here for the case of a flat space–time, this result can be generalized to a general curved space–time manifold [95].

### 5.3.4 Local Cohomology of $s$ Modulo $d$. The Russian Formula

There is a way to construct explicit solutions of the local cohomology of $s$ modulo $d$ by means of a generalized transgression formula [104, 105, 106, 107, 83, 108], also called the "Russian formula" [109]. These solutions are expressed in terms of the differential algebra generated by exterior products of the differential forms $A$ and $c$ and of their exterior derivatives. It turns out however that they give the general solution, up to elements belonging to the local cohomology of $s$: it is in particular not needed to postulate that the functional space is restricted to this differential algebra. We shall however not enter into the details [95, 96] of this statement, keeping ourselves to the derivation of the Russian formula.

The first step of the construction consists in defining a "generalized" exterior derivative

$$\tilde{d} = d + s \,, \tag{5.46}$$

which is nilpotent due to the anticommutativity of $d$ and $s$, and in introducing a one parameter family of "generalized" Yang–Mills forms, together with their associated field strengths

$$\tilde{A}_t = t(A + c) \,, \quad \tilde{F}_t = \tilde{d}\tilde{A}_t - \frac{i}{2}[\tilde{A}_t, \tilde{A}_t] \,, \quad 0 \le t \le 1 \,. \tag{5.47}$$

The BRS transformation laws (5.30) are equivalent to the "horizontality condition"

$$\tilde{F}_1 = F \,. \tag{5.48}$$

This can easily be checked by expanding both sides in powers of the ghost and by identifying terms with the same ghost number.

Given a G–invariant polynomial $P(X_1, \cdots, X_n)$ symmetric in its $n$ arguments $X_k$, $X_k$ belonging to the Lie algebra of G, we consider the generalized "homotopy formula"

$$P((\tilde{F}_1)^n) = \tilde{d}\tilde{\Omega}_{(2n-1)} \,, \tag{5.49}$$

where

$$\tilde{\Omega}_{(2n-1)} = n \int_0^1 dt\, P(A + c, (\tilde{F}_t)^{n-1}) \tag{5.50}$$

is a generalized Chern–Simons $(2n - 1)$-form[4]. Eq. (5.49) follows from the triviality [110] of the cohomology of $\tilde{d}$ on the space of local forms, and from the closure condition $\tilde{d}P = 0$ which is a consequence of the generalized Bianchi identity $\tilde{d}\tilde{F}_t - i[\tilde{A}_t, \tilde{F}_t] = 0$.

The horizontality condition (5.48) then implies the equation

$$\tilde{d}\tilde{\Omega}_{(2n-1)} = P\left((F)^n\right) \,. \tag{5.51}$$

---

[4]We use the convention

$$P(X_1, \cdots, X_p, (Y)^q) = P(X_1, \cdots, X_p, Y, \cdots, Y) \,,$$

with $q$ terms $Y$ on the right–hand side.

This is the *Russian formula*. Expanding its left–hand side according to the ghost number,

$$\tilde{\Omega}_{(Q)} = \sum_{p=0}^{Q} \Omega_p^{Q-p} , \quad \text{with } Q = 2n - 1 , \tag{5.52}$$

and identifying the terms of the same order, we get the descent equations (5.5).

*Remarks.*

1. $Q$ must be odd.

2. $Q$ may be greater than the space–time dimension $D$. All forms of degree $p > D$ appearing in the formula above have to be taken identically equal to 0.

It is an easy exercise to illustrate the Russian formula by applying it to the case of a simple Lie group, where the symmetric invariant polynomials $P$ are given by the symmetrized traces:

$$P(X_1, \cdots, X_n) = \text{Str}(X_1, \cdots, X_n) := \sum_{\text{permutations}} \frac{1}{n!} \text{Tr} \left( X_{i_1} \cdots X_{i_n} \right) ,$$

The result for $n = 3$ and $n = 2$, i.e., for $Q = 5$ and $Q = 3$, is given by the ladders (5.43) and (5.44).

## 5.4 General Solution of the Classical Descent Equations in Topological Theory

Topological field theories [111, 112, 113], which will be studied in Chap. 7, are characterized in particular by the fact that all fields and antifields[5] are grouped in complete ladders $\phi^{(i)}$ $(i = 1 \cdots I)$, according to Def. 5.1:

$$\phi_{(i)} = \left\{ \phi_{(i)p}^{Q^i-p} , \quad 0 \leq p \leq D \right\} , \tag{5.53}$$

and obey the descent equations (5.5). The meaning of the different fields, as well as the explicit expression of the coboundary operator $\delta$ are given in Chap. 7. We shall content ourselves here to deal with the operator $\delta_0$, the zero'th order term of the expansion (5.12) corresponding to the filtration operator

$$\mathcal{N} = \sum_{i=1}^{I} \sum_{p=0}^{D} \int \phi_{(i)p}^{Q^i-p} \frac{\delta}{\delta \phi_{(i)p}^{Q^i-p}} . \tag{5.54}$$

The action of $\delta_0$ reads

$$\delta_0 \phi_{(i)p}^{Q^i-p} = -d\phi_{(i)p-1}^{Q^i-p+1} , \quad 1 \leq p \leq D ,$$

$$\delta_0 \phi_{(i)0}^{Q^i} = 0 . \tag{5.55}$$

---

[5]In the terminology of Batalin–Vilkovisky [114], to each field $\phi$ corresponds an antifield $\phi^*$, which is an external field, possibly shifted, coupled to the BRS variation of $\phi$. An example is provided in Chap. 4 by the shifted external field $\hat{\rho}^\mu$ (4.24) coupled to the BRS variation of the gauge field $A_\mu$.

This is nothing else than a set of descent equations, here obeyed by the elementary fields of the theory themselves.

When discussing topological field theories in Chap. 7 we shall need the local cohomology of $\delta$ modulo $d$ (see Def. 5.5), hence that of $\delta_0$ modulo $d$ in the presence of complete ladder fields. In order to solve the local cohomology of $\delta_0$ modulo $d$, expressed by the descent equations

$$\delta_0 \Omega_p^{Q-p} + d\Omega_{p-1}^{Q-p+1} = 0 , \quad 1 \leq p \leq P ,$$
$$\delta_0 \Omega_0^Q = 0 ,$$

(5.56)

we have to know the cohomology of $\delta_0$. The latter is given by the following proposition[6]:

**Proposition 5.13** *The local cohomology of the operator $\delta_0$, whose action on the complete ladder fields (5.53) is defined by (5.55), depends on the formers only through their zero–form components $\phi_{(i)0}^{Q^i}$, undifferentiated.*

*Proof.* [89] The proof is similar to the one of Prop. 5.11. The space of local fonctionals we have to consider consists of all the polynomials in the symmetric and antisymmetric derivatives of the components $\phi_{[\mu_1 \cdots \mu_p]}^q$ $(0 \leq p \leq D)$ of the forms $\phi_{(i)p}^q$ (5.53):

$$S_{\mu_1 \cdots \mu_p, \nu_1 \cdots \nu_r}^q = \partial_{\nu_1} \cdots \partial_{(\nu_r} \phi_{\mu_1)\cdots \mu_p}^q , \quad p = 1, \cdots D; \ r \geq 0 ,$$

$$A_{\mu_1 \cdots \mu_p, \nu_1 \cdots \nu_r}^q = \partial_{\nu_1} \cdots \partial_{[\nu_r} \phi_{\mu_1]\cdots \mu_p}^q , \quad p = 1, \cdots D; \ r \geq 0 ,$$

(5.57)

$$A_{\nu_1 \cdots \nu_r}^q = \partial_{\nu_1} \cdots \partial_{\nu_r} \phi^q , \quad r \geq 1 ,$$

The action of $\delta_0$ on these derivatives reads

$$\delta_0 S_{\mu_1 \cdots \mu_p, \nu_1 \nu_2 \cdots \nu_r}^q = A_{\mu_1 \cdots \mu_{p-1}, \mu_p \nu_1 \cdots \nu_r}^{q+1} , \quad p = 1, \cdots D; \ r \geq 0 ,$$

$$\delta_0 A_{\mu_1 \cdots \mu_p, \nu_1 \cdots \nu_r}^q = 0 , \quad p = 1, \cdots D; \ r \geq 0 ,$$

(5.58)

$$\delta_0 \phi^q = 0 .$$

One sees that all fields and their derivatives – *except* the undifferentiated 0–form $\phi^q$ – are grouped in BRS doublets. It follows from Prop. 5.8 that the cohomology cannot depend on these derivatives, but only on $\phi^q$.                                     ■

The solution of the cohomology of $\delta_0$ modulo $d$ is now straightforward and is given by the following proposition:

**Proposition 5.14**

(1) *Any nontrivial ladder $\Omega$ is characterized through one 0–form $\Omega_0^Q$ belonging to the cohomology of $\delta_0$. $\Omega_0^Q$ is a $\delta_0$–invariant polynomial in the undifferentiated 0–forms $\phi_{(i)0}^{Q^i}$.*

---

[6]The formulation given here is slightly more general than needed: it covers theories which depend also on fields which are not elements of complete ladder fields.

(2) Given $\Omega_0^Q$, a nontrivial representative for the set of the higher elements of the ladder $\Omega$ is given by the expressions

$$\Omega_p^{Q-p} = \frac{1}{p!} \nabla^p \Omega_0^Q , \tag{5.59}$$

where the operator $\nabla$ is defined by its action on the elementary ladder fields:

$$\nabla \phi_{(i)\,p}^{Q^i - p} = (p+1) \phi_{(i)\,p+1}^{Q^i - p - 1} , \quad p = 0, \cdots, D-1 ,$$

$$\nabla \phi_{(i)\,D}^{Q^i - D} = 0 . \tag{5.60}$$

*Proof.* The first statement follows from the observation that, due to Prop. 5.13, it is only while solving the descent equation for $\Omega_0^Q$ (the last of Eqs. (5.56)) that we have to deal with a nontrivial cohomology of $\delta_0$. The second statement is a consequence of the commutation relations

$$[\delta_0, \nabla] = -d , \quad [d, \nabla] = 0 . \tag{5.61}$$

Indeed:

$$\delta_0 \frac{1}{p!} \nabla^p \Omega_0^Q = \frac{1}{p!} [\delta_0, \nabla^p] \, \Omega_0^Q = -\frac{1}{(p-1)!} d \nabla^{p-1} \Omega_0^Q .$$

■

*Remarks.*

1. Using the generalized form notation (c.f. (5.52))

$$\tilde{\Omega}_{(Q)} = \sum_{p=0}^{D} \Omega_p^{Q-p} = \sum_{p=0}^{D} \frac{1}{p!} \nabla^p \Omega_0^Q , \tag{5.62}$$

we can write

$$\tilde{\Omega}_{(Q)} = e^{\nabla} \Omega_0^Q \left( \phi_{(i)0}^{Q^i} \right) = \Omega_0^Q \left( \tilde{\phi}_{(i)0}^{Q^i} \right) , \tag{5.63}$$

where

$$\tilde{\phi}_{(i)0}^{Q^i} = \sum_{p=0}^{D} \phi_{(i)p}^{Q^i - p} . \tag{5.64}$$

(The exponential series terminates with the term of maximum degree $p = D$).

This way of expressing the result (5.59) points out its equivalence with the results and the notations of [115] in the context of the topological $BF$ systems (see Chap. 7).

2. In the examples which will be considered later on the decomposition (5.61) turns out to hold for the full coboundary operator $\delta$.

3. The decomposition (5.61) of the exterior derivative $d$ and its consequences generalize to Yang–Mills and gravitation theories [116, 117].

## 5.5 Renormalized Descent Equations in Yang–Mills Theory

Let us consider a Yang–Mills theory of the type studied in Chap. 4. We shall restrict our attention to the four–dimensional case, i.e., to the case where the theory is power–counting renormalizable. Our aim is to show the possibility of constructing a quantum extension

$$\Omega \cdot \Gamma = \left\{ \Omega_p^{Q-p} \cdot \Gamma \ , \ 0 \le p \le P \le 4 \ , \ Q \ \text{fixed} \right\} \tag{5.65}$$

of the classical ladder $\Omega$ of composite fields (5.26) obeying the quantum descent equations

$$
\begin{aligned}
&\mathcal{S}_\Gamma \left[ \Omega_p^{Q-p} \cdot \Gamma \right] + d \left[ \Omega_{p-1}^{Q-p+1} \cdot \Gamma \right] = 0 \ , \quad 4 \ge p \ge 1 \ , \\
&\mathcal{S}_\Gamma \left[ \Omega_0^Q \cdot \Gamma \right] = 0 \ ,
\end{aligned}
\tag{5.66}
$$

where $\mathcal{S}_\Gamma$ is the linearized Slavnov–Taylor operator (4.78). We shall assume this ladder to be complete, i.e., $P = 4$. Although the proof we are going to give concerns explicitly the special case of total degree $Q = 5$, for a simple Lie group, it can be generalized to any $Q$ and a general compact Lie group.

### 5.5.1 Renormalization

Our first task will be to introduce external fields coupled to the elements of $\Omega$. These external fields constitute a complete ladder of $(4 - p)$–forms

$$u = \left\{ u_{4-p}^{p-Q} \ , \quad 0 \le p \le 4 \right\} \ , \tag{5.67}$$

transforming under BRS as

$$
\begin{aligned}
&s u_{4-p}^{p-Q} + d u_{3-p}^{p-Q+1} = 0 \ , \quad 0 \le p \le 3 \ , \\
&s u_0^{4-Q} = 0 \ .
\end{aligned}
\tag{5.68}
$$

The coupling of $u$ with $\Omega$ is provided by the addition of the term

$$S_{(u)} = \int \sum_{p=0}^{4} u_{4-p}^{p-Q} \Omega_p^{Q-p} \ , \tag{5.69}$$

to the classical action (4.14). The BRS invariance of this action is equivalent to the fulfillment of the classical descent equations by the classical ladder $\Omega$.

The corresponding quantum theory will be characterized by the new Slavnov–Taylor operator

$$
\begin{aligned}
\mathcal{S}(\Gamma) := \int d^4 x \left( \text{Tr} \, \frac{\delta \Gamma}{\delta \rho^\mu} \frac{\delta \Gamma}{\delta A_\mu} + \frac{\delta \Gamma}{\delta \bar{Y}} \frac{\delta \Gamma}{\delta \psi} - \frac{\delta \Gamma}{\delta Y} \frac{\delta \Gamma}{\delta \bar{\psi}} + \text{Tr} \, \frac{\delta \Gamma}{\delta \sigma} \frac{\delta \Gamma}{\delta c} + \text{Tr} \, B \frac{\delta \Gamma}{\delta \bar{c}} \right) \\
- \sum_{p=0}^{3} \int \left( d u_{3-p}^{p-Q+1} \frac{\delta \Gamma}{\delta u_{4-p}^{p-Q}} \right) + \chi \frac{\partial \Gamma}{\partial \alpha} \ ,
\end{aligned}
\tag{5.70}
$$

and its linearized form

$$\mathcal{S}_\Gamma := \int d^4x \left( \mathrm{Tr} \, \frac{\delta\Gamma}{\delta\rho^\mu} \frac{\delta}{\delta A_\mu} + \frac{\delta\Gamma}{\delta A_\mu} \frac{\delta}{\delta\rho^\mu} + \frac{\delta\Gamma}{\delta\bar{Y}} \frac{\delta}{\delta\psi} + \frac{\delta\Gamma}{\delta\psi} \frac{\delta}{\delta\bar{Y}} \right.$$
$$- \frac{\delta\Gamma}{\delta Y} \frac{\delta}{\delta\bar{\psi}} - \frac{\delta\Gamma}{\delta\bar{\psi}} \frac{\delta}{\delta Y} + \mathrm{Tr} \, \frac{\delta\Gamma}{\delta\sigma} \frac{\delta}{\delta c} + \mathrm{Tr} \, \frac{\delta\Gamma}{\delta c} \frac{\delta}{\delta\sigma} \qquad (5.71)$$
$$\left. + \mathrm{Tr} \, B \frac{\delta}{\delta\bar{c}} \right) - \sum_{p=0}^{3} \int \left( du_{3-p}^{p-Q+1} \frac{\delta}{\delta u_{4-p}^{p-Q}} \right) + \chi \frac{\partial}{\partial\alpha} \ ,$$

which takes into account the BRS transformations (5.68) of the external forms $u$. We have also considered in (5.70) the BRS variation $\chi$ of the gauge parameter $\alpha$ in order to control the gauge (in)dependence – as explained in Sect. 4.5 – of the quantum ladder.

The construction of the latter will be achieved by showing that the $u$–dependent vertex functional $\Gamma$ obeys the new Slavnov–Taylor identity

$$\mathcal{S}(\Gamma) = 0 \ . \qquad (5.72)$$

Indeed, the quantum ladder being defined by

$$\Omega_p^{Q-p} \cdot \Gamma = \left. \frac{\delta\Gamma}{\delta u_{4-p}^{p-Q}} \right|_{u=0} , \qquad (5.73)$$

differentiation of the Slavnov–Taylor identity with respect to $u_{4-p}^{p-Q}$ will yield the descent equation (5.66) at level $p$.

As we have discussed at length in Chap. 4, the validity of the new Slavnov identity is equivalent to the absence of anomalies, i.e., to the vanishing of the cohomology in the space of integrated local functionals with ghost number 1. Assuming the absence of the usual gauge anomaly, we are left to consider the terms which depend on $u$. We have thus to study the cohomology of $s$ modulo $d$ in the sector of the $u$–dependent local functionals $\Omega_p^{Q-p}$.

We shall now show that this cohomology is empty, specializing to the case $Q = 5$, i.e., to the ladder (5.43) corresponding to the gauge anomaly[7], and restricting the gauge group to be a simple Lie group,

The external field ladder $u$ being then complete, Prop. 5.13 (with $\delta_0$ replaced by $s$) applies: the local cohomology of $s$ depend on $u$ only through its undifferentiated 0–form component $u_0^{-1}$. Moreover Prop. 5.12 of Subsect. 5.3.2 implies that this cohomology will depend on $c$ and $A$ only through the invariants $\mathrm{Tr}\, c^n$ and those made with the Yang–Mills curvature $F$ and its derivatives. In the sector of form degree $p$ and ghost number $5 - p$, with $p \leq 4$, we are considering here, the only candidates having a dimension compatible with power–counting (i.e., bounded by $p$) would be the 0–forms $u_0^{-1}\mathrm{Tr}\, c^6$ and $u_0^{-1}(\mathrm{Tr}\, c^3)^2$, which vanish due to the cyclicity of the trace and the fermionic nature of the ghost fields.

The local cohomology of $s$ thus vanishing in the $u$–dependent sector, the local cohomology of $s$ modulo $d$ will vanish, too, as a consequence of Prop. 5.4. This means in particular the absence of $u$–dependent anomaly for the Slavnov–Taylor identity (5.72). This ends the proof of the absence of anomaly in the renormalization of the ladder (5.43).

---

[7]The case $Q = 3$, which corresponds to the U(1) anomaly, is treated in [102].

### 5.5.2 The Callan–Symanzik Equation for the Ladder of the Gauge Anomaly

The stability of the renormalization of the ladder considered in the last subsection corresponds to the search of the $u$–dependent BRS invariant counterterms which can be freely added to the action at each order. This amounts to solve the cohomology of $s$ modulo $d$, as for the anomaly, but now in the sector of total degree $Q = 4$. The relevant ladder is given by (5.26) for $P = D = 4$ and $Q = 4$. It is again determined by the local cohomology of $s$. The only nontrivial $u$–dependent element of the latter – power–counting constraints being taken into account – is the 0–form of ghost number 4

$$\Omega_0^4 = -\frac{1}{10} u_0^{-1} \text{Tr} \, c^5 \, , \tag{5.74}$$

as a consequence of Prop. 5.12. This leads to a ladder which is unique up to the equivalence defined in Def. 5.2. Its 4–form element leads to the invariant counterterm, unique up to a BRS variation,

$$L = \int \Omega_4^0 = \int \sum_{p=0}^{4} u_{4-p}^{p-5} \Omega_p^{5-p} \, , \tag{5.75}$$

where the $\Omega_p^{5-p}$ are the elements of the gauge anomaly ladder (5.43). This counterterm being identical to the $u$–dependent term $S_{(u)}$ of the classical action (c.f. (5.69)), one can write it in the form

$$L = \mathcal{N}_u S \, , \qquad \text{with} \quad \mathcal{N}_u = \int \sum_{p=0}^{4} u_{4-p}^{p-5} \frac{\delta}{\delta u_{4-p}^{p-5}} \, , \tag{5.76}$$

where $S$ denotes now the classical action (4.14) with the inclusion of the $u$–dependent term (5.69). The most general $u$–dependent $s$–invariant counterterm is thus given by the linear superposition of (5.76) and of the BRS variation $b\hat{\Delta}$ of an arbitrary classical insertion $\hat{\Delta}$ of dimension 4 and ghost number $-1$. Here

$$b = \mathcal{S}_S$$

is the linearized Slavnov–Taylor operator (5.71), taken for the the classical action $S$.

At the quantum level these considerations lead, through a way which by now should be familiar to the reader (c.f. Sect. 3.4.2, 3.5 and Subsect. 4.4.3), to the following Callan–Symanzik for the $u$–dependent theory:

$$\mathcal{C}_{(u)} \Gamma := (\mathcal{C} + \gamma_u \mathcal{N}_u) \Gamma = \mathcal{S}_\Gamma \left[ \hat{\Delta} \cdot \Gamma \right] \, , \tag{5.77}$$

where $\mathcal{C}$ is the Yang–Mills Callan–Symanzik operator defined by (4.66). We don't specify more precisely the right–hand side, which corresponds to nonphysical renormalizations since it is cohomologically trivial. In contrast, the cohomologically nontrivial $u$–anomalous dimension term in the left–hand side corresponds to the renormalization of the ladder of the gauge anomaly – a physically meaningful object. We shall show in Chap. 6, Sect. 6.3, that the "anomalous dimension" $\gamma_u$ of this ladder actually vanishes.

An important observation is that the coefficient of the cohomologically nontrivial counterterm (5.76) must be independent of the gauge parameter $\alpha$, since any BRS–invariant function of the gauge parameter is trivial as we have shown in Sect. 4.5. The direct consequence for the renormalized theory is the gauge independence of the anomalous dimension $\gamma_u$:

$$\frac{\partial}{\partial \alpha} \gamma_u = 0 \ . \tag{5.78}$$

Finally, differentiating (5.77) with respect to $u_{4-p}^{p-5}$ we obtain the Callan–Symanzik equation for the renormalized ladder:

$$(\mathcal{C} + \gamma_u) \left[ \Omega_p^{5-p} \cdot \Gamma \right] = \left( \mathcal{S}_\Gamma \frac{\delta}{\delta u_{4-p}^{p-5}} \left[ \hat{\Delta} \cdot \Gamma \right] \right) \Bigg|_{u=0} \ , \quad p = 0, \cdots, 4 \ . \tag{5.79}$$

# 6. Nonrenormalization Theorems

In Chap. 4, we have seen that the BRS invariance of gauge theories may be broken at the quantum level by the gauge anomaly, which would render the theory inconsistent. However our purely algebraic approach, based on the cohomology in the space of local field polynomials, gives only a potential anomaly. It does not yield the value of its coefficient $r$ (c.f. (4.55)), which is needed in order to know if the anomaly is actually present. This coefficient $r$ is a power series in the coupling constants or in $\hbar$, and a *priori* receives contributions from Feynman diagrams of all orders. But a remarkable nonrenormalization theorem due to Adler and Bardeen [118] states that the gauge anomaly is definitively absent if the coefficient $r$ vanishes at the first order in $\hbar$. This means that, for checking the consistency of a physical model, it is sufficient to examine the one–loop contribution to the anomaly.

The aim of this chapter is to present a proof of the nonrenormalization theorem using the techniques of algebraic renormalization. Owing to the gauge independence of the coefficient $r$ proven at the end of Sect. 4.5, one can specialize to a particular gauge. We shall choose the Landau gauge which, as we shall see, possesses peculiar ultraviolet finiteness properties thanks to an identity, the "antighost equation", which controls the coupling of the Faddeev–Popov ghost $c$ in such a way that the latter has no anomalous dimension.

For simplicity we shall consider here a gauge model of massless chiral fermions only, and we shall assume that the gauge group is simple. However the validity of the theorem persists in the case of a general compact Lie group and of models including scalar fields and Yukawa couplings [87, 88].

## 6.1 The Antighost Equation

As said above, the Landau gauge has very special features as compared to a generic linear gauge (cf. (4.8)). This is due to the existence, besides the Slavnov–Taylor identity (4.12), of a further identity, the antighost equation [119], which controls the dependence of the theory on the ghost $c$. In particular this equation implies that the ghost field $c$ and the composite field cocycles $\mathrm{Tr}\, c^{(2p+1)}$ entering the descent equations (5.27) have vanishing anomalous dimension. This result, as we shall see in detail, allows to give an algebraic proof of the Adler–Bardeen nonrenormalization theorem [118] for the gauge anomaly (4.55).

We have met the antighost equation in Chap. 4 (see (4.99)) in the context of a cohomology constrained by a rigid symmetry. We shall see now that its validity is quite general – in the Landau gauge.

Before deriving the antighost equation let us recall some basic properties concerning the Landau gauge. The latter is defined by the vanishing of the gauge parameter $\alpha$ of (4.8), i.e., by the condition

$$\frac{\delta S}{\delta B} = \partial^\mu A_\mu , \tag{6.1}$$

which is easily seen to imply that the propagator of the gauge field $A$ is transverse. Indeed, in terms of the connected functional $Z^{c(0)}$ of the tree approximation, the gauge condition (6.1) becomes

$$\partial^\mu \frac{\delta Z^{c(0)}}{\delta J_A^\mu} = -J_B , \tag{6.2}$$

and, differentiating with respect to $J_A^\nu$ and setting all the sources equal to zero, one gets

$$\partial_x^\mu \frac{\delta^2 Z^{c(0)}}{\delta J_A^\mu(x)\delta J_A^\nu(y)}\bigg|_{J=0} = 0 , \tag{6.3}$$

which expresses the transversality of the two–point gauge function.

Due to the condition (6.1) the complete classical action (4.14) takes now the form

$$S = S_{\text{inv}} + S_{\text{gf}} + S_{\text{ext}} , \tag{6.4}$$

with $S_{\text{inv}}$ and $S_{\text{ext}}$ as in (4.7) and (4.11) and $S_{\text{gf}}$ given by

$$
\begin{aligned}
S_{\text{gf}} &= s\,\mathrm{Tr} \int d^4x \; \bar{c}\partial^\mu A_\mu \\
&= \mathrm{Tr} \int d^4x \; (B\partial^\mu A_\mu - \bar{c}\partial^\mu(\partial_\mu c + i[c, A_\mu])) .
\end{aligned} \tag{6.5}
$$

In order to derive the antighost equation let us differentiate the action (6.4) with respect to the ghost field $c$,

$$\frac{\delta S}{\delta c} = \partial^2 \bar{c} + \partial\rho + i[\partial^\mu \bar{c}, A_\mu] + i[\rho^\mu, A_\mu] - i[\sigma, c] + i(\bar{Y}T^a\psi + \bar{\psi}T^aY)\tau_a . \tag{6.6}$$

Integrating on space–time and using the Landau gauge condition (6.1) we get the antighost equation we were looking for:

$$\bar{\mathcal{G}}S = \Delta_{\text{cl}}^G , \tag{6.7}$$

where

$$\bar{\mathcal{G}} = \int d^4x \left( -i\frac{\delta}{\delta c} + \left[\bar{c}, \frac{\delta}{\delta B}\right] \right) , \tag{6.8}$$

and

$$\Delta_{\text{cl}}^G = \int d^4x \left( [\rho^\mu, A_\mu] - [\sigma, c] + (\bar{Y}T^a\psi + \bar{\psi}T^aY)\tau_a \right) . \tag{6.9}$$

Notice that the breaking $\Delta_{\text{cl}}^G$, being linear in the quantum fields $A_\mu$, $\psi$, $\bar{\psi}$, will not be renormalized, i.e., it will remain a classical breaking. Equation (6.7) has then a form which allows us to try to extend it as it stands to the quantum level. That such an extension always holds will be proven in the next subsection.

*Remark.* In the case of a general linear covariant gauge (cf. (4.8)) the right–hand side of (6.7) would no more be linear in the quantum fields, due to the presence of the term $\alpha \int d^4x \, [\bar{c}, B]$. The latter, being quadratic in the fields, would be subject to renormalization, i.e., it would have to be defined as an insertion. This would spoil the usefulness of the antighost equation. Therefore we keep the Landau gauge $\alpha = 0$.

The antighost operator $\bar{\mathcal{G}}$ forms together with the Slavnov–Taylor (4.12), gauge condition (6.1) and ghost equation (4.20) operators, an algebra whose most important relations, denoting with $\mathcal{F}$ an arbitrary functional with even ghost number, are:

$$(i) \quad \bar{\mathcal{G}}\mathcal{S}(\mathcal{F}) + \mathcal{S}_{\mathcal{F}}(\bar{\mathcal{G}}\mathcal{F} - \Delta^G_{cl}) = \mathcal{R}^{rig}\mathcal{F} \ ,$$

$$(ii) \quad \frac{\delta}{\delta B}(\bar{\mathcal{G}}\mathcal{F} - \Delta^G_{cl}) - \bar{\mathcal{G}}(\frac{\delta \mathcal{F}}{\delta B} - \partial A) = 0 \ ,$$

$$(iii) \quad \bar{\mathcal{G}}_a \mathcal{G}_b \mathcal{F} + \mathcal{G}_b(\bar{\mathcal{G}}_a \mathcal{F} - \Delta^G_{a,cl}) = i f_{abc}\left(\frac{\delta \mathcal{F}}{\delta B_c} - \partial A^c\right) \ ,$$

$$(iv) \quad \bar{\mathcal{G}}_a\left(\bar{\mathcal{G}}_b \mathcal{F} - \Delta^G_{b,cl}\right) + \bar{\mathcal{G}}_b\left(\bar{\mathcal{G}}_a \mathcal{F} - \Delta^G_{a,cl}\right) = 0 \ ,$$

$$(6.10)$$

where

$$\mathcal{G}_a = \mathrm{Tr}\,(\tau_a \mathcal{G}) \ , \qquad \bar{\mathcal{G}}_a = \mathrm{Tr}\,(\tau_a \bar{\mathcal{G}}) \ , \tag{6.11}$$

and $\mathcal{S}_{\mathcal{F}}$ denotes the complete linearized Slavnov–Taylor operator (4.22)

$$\mathcal{S}_{\mathcal{F}} = \int d^4x \left( \mathrm{Tr}\frac{\delta \mathcal{F}}{\delta \rho^\mu}\frac{\delta}{\delta A_\mu} + \mathrm{Tr}\frac{\delta \mathcal{F}}{\delta A_\mu}\frac{\delta}{\delta \rho^\mu} + \frac{\delta \mathcal{F}}{\delta \bar{Y}}\frac{\delta}{\delta \psi} + \frac{\delta \mathcal{F}}{\delta \psi}\frac{\delta}{\delta \bar{Y}} - \frac{\delta \mathcal{F}}{\delta Y}\frac{\delta}{\delta \bar{\psi}} \right.$$
$$\left. - \frac{\delta \mathcal{F}}{\delta \bar{\psi}}\frac{\delta}{\delta Y} + \mathrm{Tr}\frac{\delta \mathcal{F}}{\delta \sigma}\frac{\delta}{\delta c} + \mathrm{Tr}\frac{\delta \mathcal{F}}{\delta c}\frac{\delta}{\delta \sigma} + \mathrm{Tr}\,B\frac{\delta}{\delta \bar{c}} \right) \ , \tag{6.12}$$

which obeys the identity

$$\mathcal{S}_{\mathcal{F}}\mathcal{S}(\mathcal{F}) = 0 \ . \tag{6.13}$$

The operator $\mathcal{R}^{rig}$ in the first of the equations (6.10) is the Ward operator for the rigid gauge transformations

$$\mathcal{R}^{rig}\mathcal{F} = \int d^4x \left( \left[A, \frac{\delta \mathcal{F}}{\delta A}\right] + \left[B, \frac{\delta \mathcal{F}}{\delta B}\right] + \left[\sigma, \frac{\delta \mathcal{F}}{\delta \sigma}\right] + \left\{c, \frac{\delta \mathcal{F}}{\delta c}\right\} + \left\{\bar{c}, \frac{\delta \mathcal{F}}{\delta \bar{c}}\right\} \right.$$
$$\left. + \left\{\rho, \frac{\delta \mathcal{F}}{\delta \rho}\right\} - \left(\frac{\delta \mathcal{F}}{\delta \psi}T^a\psi + \bar{Y}T^a\frac{\delta \mathcal{F}}{\delta \bar{Y}} + \bar{\psi}T^a\frac{\delta \mathcal{F}}{\delta \bar{\psi}} - \frac{\delta \mathcal{F}}{\delta Y}T^aY \right)\tau_a \right) \ , \tag{6.14}$$

which leaves invariant the classical action (6.4):

$$\mathcal{R}^{rig}\mathcal{S} = 0 \ . \tag{6.15}$$

One sees from (6.10) $(i)$ that, in the Landau gauge, the rigid gauge invariance results from the Slavnov–Taylor identity and the antighost equation.

### 6.1.1 Renormalization of the Antighost Equation

We shall now prove that the antighost equation (6.7) can be extended to all orders of perturbation theory. The proof we are going to give doesn't make use of the Slavnov–Taylor identity (4.12), it relies only on the assumption of the gauge condition (6.1) and of the ghost equation (4.20). We shall be able then to use the antighost equation also in the case in which the Slavnov–Taylor identity turns out to be anomalous.

Let us write a broken antighost equation

$$\bar{\mathcal{G}}_a \Gamma = \Delta^G_{a,\text{cl}} + \hbar^q \Xi_a \cdot \Gamma , \qquad q \geq 1$$
$$\Xi_a \cdot \Gamma = \Xi_a + O(\hbar) ,$$
(6.16)

where $\Xi_a$ represents the possible breaking induced by the radiative corrections. According to the Quantum Action Principle the lowest–order nonvanishing contribution to the breaking – of order $\hbar$ at least – is an integrated local functional with dimension four and ghost number $-1$ which, due to relations $(ii)$, $(iii)$ and $(iv)$ of (6.10), is constrained by

$$\frac{\delta \Xi_a}{\delta B_c} = 0 , \qquad \mathcal{G}_c \Xi_a = 0 ,$$
(6.17)

$$\bar{\mathcal{G}}_a \Xi_b + \bar{\mathcal{G}}_b \Xi_a = 0 .$$
(6.18)

The antighost equation (6.7) will extend to the quantum level if one is able to prove that the breaking $\Xi_a$ can be written as a $\bar{\mathcal{G}}$–variation, i.e.,:

$$\Xi_a = \bar{\mathcal{G}}_a \hat{\Xi} ,$$
(6.19)

for some local functional $\hat{\Xi}$, the latter being then absorbed as a counterterm. Conditions (6.17) imply that $\Xi_a$ is $B$–independent and that the fields $\bar{c}$ and $\rho$ enter only through the combination (4.24). It follows that $\Xi_a$ can be parametrized as

$$\Xi_a = \int d^4x \left( \omega_{abc} c^b \sigma^c + \nu_{abc} \hat{\rho}^{b\mu} A^c_\mu + \xi_a{}^b (\bar{Y} T_b \psi) + \beta_a{}^b (\bar{\psi} T_b Y) \right) ,$$
(6.20)

where $\omega_{abc}, \nu_{abc}, \xi_a{}^b, \beta_a{}^b$ are arbitrary coefficients.

Since

$$\int d^4x \, \nu_{abc} \hat{\rho}^{b\mu} A^c_\mu = i \bar{\mathcal{G}}_a \int d^4x \, \nu_{dbc} c^d \hat{\rho}^{b\mu} A^c_\mu ,$$
(6.21)

and since

$$\int d^4x \left( \xi_a{}^b (\bar{Y} T_b \psi) + \beta_a{}^b (\bar{\psi} T_b Y) \right) = i \bar{\mathcal{G}}_a \int d^4x \left( \xi_b{}^c c^b (\bar{Y} T_c \psi) + \beta_b{}^c c^b (\bar{\psi} T_c Y) \right) ,$$
(6.22)

the nontrivial part of $\Xi_a$ reduces to

$$\Xi_a = \int d^4x \, \omega_{abc} \sigma^b c^c .$$
(6.23)

Finally, from (6.18), it turns out that $\omega_{abc}$ is antisymmetric in the indices $a, b$

$$\omega_{abc} = -\omega_{bac}$$
(6.24)

so that

$$\int d^4x \, \omega_{abc} c^b \sigma^c = i \bar{\mathcal{G}}_a \int d^4x \, \frac{1}{2} \omega_{bdc} c^b c^d \sigma^c .$$
(6.25)

One sees thus that all the terms of (6.20) can be written as $\bar{\mathcal{G}}$–variations, meaning that the breaking $\Xi_a$ can be compensated by the introduction of suitable counterterms. We have thus proven the antighost equation at the order considered, hence to all orders by induction.

*Remark.* Notice that in the proof of the renormalizability of (6.7) no use has been made of the rigid invariance (6.15). The validity of the antighost equation will persist then in a more general context, as in the case of a spontaneously broken gauge invariance.

## 6.2 Nonrenormalization Theorems

Having discussed the renormalizability of the antighost equation (6.7), let us turn to the study of some important consequences implied by this identity.

In what follows and in view of our aim to use the antighost equation in the proof of the Adler–Bardeen nonrenormalization theorem [118], we shall assume to be in the general case in which the Slavnov–Taylor identity may be broken by the gauge anomaly. The theory is described then by a vertex functional $\Gamma$ which is required to obey:

(1) the Landau gauge–fixing condition (6.1) and the ghost equation (4.20)

$$\frac{\delta\Gamma}{\delta B} = \partial^\mu A_\mu \,, \qquad \mathcal{G}\Gamma = 0 \,, \tag{6.26}$$

(2) the antighost equation (6.7) and the rigid gauge invariance (6.15)

$$\bar{\mathcal{G}}\Gamma = \Delta_{\text{cl}}^G \,, \qquad \mathcal{R}^{\text{rig}}\Gamma = 0 \,, \tag{6.27}$$

(3) the Slavnov–Taylor identity, anomalous at the order $n$,

$$\mathcal{S}(\Gamma) = r\hbar^n \mathcal{A} + O(\hbar^{n+1}) \tag{6.28}$$

with $\mathcal{A}$ given by (4.55). In particular (6.28) implies that the gauge invariance can be maintained only up to the order $\hbar^{n-1}$. As a consequence the results we shall establish will be Slavnov–Taylor invariant only up to this order. Moreover, they will extend to all orders in the case in which the Slavnov–Taylor identity turns out to be anomaly free. The aim of the present chapter is just to show that this is always possible if one requires the gauge anomaly to be absent in the one–loop approximation.

### 6.2.1 Nonrenormalization of the Ghost Field

In order to prove the nonrenormalization of the ghost field $c$ let us recall that, due to the conditions (6.26), (6.28), the BRS invariant counterterms $L_{(i)}$ which one can freely add to the action up to the order $\hbar^{n-1}$ are $B$–independent, depend on $\bar{c}$ and $\rho$ only through the combination (4.24) and satisfy the condition (4.56)

$$bL_{(i)} = 0 \,, \tag{6.29}$$

where $b$ stands for the linearized Slavnov–Taylor operator (6.12) corresponding to the classical action:

$$b = \mathcal{S}_S \,. \tag{6.30}$$

As previously discussed (cf. Sect. 4.4), the various invariant counterterms correspond to coupling constant and field amplitude renormalizations. The particular one which corresponds to the renormalization of the ghost field $c$ is given by

$$L_{(\sigma)} = b \int d^4x \, \text{Tr} \, (\sigma c) = \int d^4x \, \text{Tr} \left( -c \frac{\delta S}{\delta c} + \sigma \frac{\delta S}{\delta \sigma} \right) . \qquad (6.31)$$

However, preservation of the antighost equation (6.27) requires that the counterterms $L_{(i)}$ have to obey the homogeneous condition

$$\bar{\mathcal{G}} L_{(i)} = 0 . \qquad (6.32)$$

It is apparent that this requirement excludes the counterterm $L_{(\sigma)}$. This also implies the absence of the counting operator

$$\mathcal{N}_c = \int d^4x \, \text{Tr} \left( c \frac{\delta}{\delta c} - \sigma \frac{\delta}{\delta \sigma} \right) \qquad (6.33)$$

in the Callan–Symanzik equation (4.66) below the order $n$. Hence the latter, if we also take into account the absence of the gauge parameter $\alpha$ in the Landau gauge, reads

$$\mathcal{C}\Gamma = \left( \mu \frac{\partial}{\partial \mu} + \hbar \beta_g \frac{\partial}{\partial g} - \hbar \gamma_A \mathcal{N}_A - \hbar \gamma_\psi \mathcal{N}_\psi \right) \Gamma = \hbar^n \Delta_c^n + O(\hbar^{n+1}) , \qquad (6.34)$$

where we have kept the factor $\hbar$ explicitly according to the fact that the first nonvanishing term of the insertion $\partial \Gamma / \partial \mu$ is at least of order $\hbar$, $\mu$ being the normalization point[1]. $\mathcal{N}_A$ and $\mathcal{N}_\psi$ are the counting operators (4.59)

$$\mathcal{N}_A = \int d^4x \, \text{Tr} \left( A \frac{\delta}{\delta A} - B \frac{\delta}{\delta B} - \bar{c} \frac{\delta}{\delta \bar{c}} - \rho \frac{\delta}{\delta \rho} \right) , \qquad (6.35)$$

$$\mathcal{N}_\psi = \int d^4x \left( \psi \frac{\delta}{\delta \psi} - \bar{Y} \frac{\delta}{\delta \bar{Y}} \right) + \left( \bar{\psi} \frac{\delta}{\delta \bar{\psi}} - Y \frac{\delta}{\delta Y} \right) , \qquad (6.36)$$

and $\Delta_c^n$ is a local integrated functional of dimension four and ghost number zero which accounts for the lack of the BRS invariance at the order $\hbar^n$ according to (6.28). Thus the antighost equation implies the vanishing of the anomalous dimension of the ghost field $c$, i.e., it doesn't suffer any renormalization. This result extends to all orders if the gauge anomaly is absent.

Let us conclude this section with an important remark concerning the Callan–Symanzik equation (6.34). Using the exact algebraic relation

$$\mathcal{CS}(\Gamma) = \mathcal{S}_\Gamma \mathcal{C}\Gamma \qquad (6.37)$$

it follows that, applying (6.34) to the anomalous Slavnov–Taylor identity (6.28) and retaining the lowest order in $\hbar$ (i.e., $\hbar^n$), one gets

$$b \Delta_c^n = \mu \frac{\partial r}{\partial \mu} \mathcal{A} \qquad (6.38)$$

---

[1] The theory considered here is massless.

which, due to the fact that the gauge anomaly $\mathcal{A}$ cannot be written as the $b$–variation of a local functional, implies that

$$\frac{\partial r}{\partial \mu} = 0 \ , \tag{6.39}$$

and

$$b\Delta_c^n = 0 \ . \tag{6.40}$$

Conditions (6.39), (6.40) show that the coefficient $r$ is independent from the renormalization point $\mu^2$ and that $\Delta_c^n$ is BRS invariant. Moreover, taking into account that $\Delta_c^n$ satisfies also the homogeneous antighost condition (6.32) and that, due to (6.26), it is $B$–independent and obeys the ghost equation, we can expand it in terms of the elements of the basis

$$\left( \frac{\partial S}{\partial g}, \ \mathcal{N}_A S, \ \mathcal{N}_\psi S \right) \ . \tag{6.41}$$

As a consequence (6.34) becomes

$$\mathcal{C}\Gamma = \left( \mu \frac{\partial}{\partial \mu} + \hbar \beta_g \frac{\partial}{\partial g} - \hbar \gamma_A \mathcal{N}_A - \hbar \gamma_\psi \mathcal{N}_\psi \right) \Gamma = \hbar^{n+1} \Delta_c^{n+1} + O(\hbar^{n+2}) \ , \tag{6.42}$$

with $\Delta_c^{n+1}$ a local integrated functional of dimension four and ghost number zero. One sees thus that the Callan–Symanzik equation, in spite of the presence of the gauge anomaly, extends in a BRS invariant way up to the order $\hbar^n$.

### 6.2.2 Nonrenormalization of $\mathrm{Tr}\, c^{(2p+1)}$

Another consequence of the antighost equation (6.7) is the ultraviolet finiteness, i.e., the vanishing of their anomalous dimension, of the composite ghost cocycles $\mathrm{Tr}\, c^{(2p+1)}$. This result is of particular importance since these cocycles are directly related, through the descent equations (5.27), to anomalies and observables. Indeed, as shown in Chap. 5, to any given set of classical field polynomials which are solutions of (5.27) corresponds a ladder of quantum insertions (5.65) which fulfil the quantum version (5.66) of the descent equations. However, BRS invariance implies that these quantum insertions have the same anomalous dimension. Therefore the vanishing of the anomalous dimension of the cocycles $\mathrm{Tr}\, c^{(2p+1)}$ implies the vanishing of the anomalous dimension for all the quantum insertions belonging to the same ladder. This property will be very useful in the analysis of the nonrenormalization theorem for the gauge anomaly.

We shall limit here the discussion to an explicit example: the proof of the ultraviolet finiteness of the ghost cocycle in the case $p = 1$, but in a way suitable for generalization to any $p$, as it has been done in [89]. Let us mention also that the cocycle $\mathrm{Tr}\, c^3$ plays a key role in the analysis of the $U(1)$ axial anomaly. Its ultraviolet finiteness is at the root of the nonrenormalization theorem for this anomaly [102].

In order to define the BRS invariant composite field cocycle

---

[2]This result can be generalized also to the case in which massive fields are present: $r$ doesn't depend on any massive parameter [88].

$$\operatorname{Tr} \frac{c^3}{3} = i f_{abc} \frac{c^a c^b c^c}{6} \, , \tag{6.43}$$

we couple it to an invariant external field $\eta$ of dimension four and ghost number $-3$, i.e., we add to (6.4) the term

$$S_{(\eta)} = -i \int d^4x \, \eta \operatorname{Tr} \frac{c^3}{3} \, , \tag{6.44}$$

so that the new complete classical action reads

$$S = S_{\text{inv}} + S_{\text{gf}} + S_{\text{ext}} + S_{(\eta)} \, . \tag{6.45}$$

From

$$\int d^4x \, \frac{\delta S_\eta}{\delta c} = i \int d^4x \, \eta c^2 = \int d^4x \, \eta \frac{\delta S}{\delta \sigma} \, , \tag{6.46}$$

it follows that the antighost equation (6.7) still holds with the same classical breaking $\Delta^G_{\text{cl}}$ but with a modified differential operator

$$\bar{\mathcal{G}} = \int d^4x \left( -i \frac{\delta}{\delta c} + \left[ \bar{c}, \frac{\delta}{\delta B} \right] + i \eta \frac{\delta}{\delta \sigma} \right) \, . \tag{6.47}$$

The gauge condition (6.1), the ghost equation (4.20) and the classical Slavnov–Taylor identity (4.12) as well as the algebraic relations (6.10) remain unchanged[3].

Before proving the ultraviolet finiteness of the ghost cocycle (6.43) let us show that the introduction of the external field $\eta$ does not affect the gauge anomaly (4.55). It is easy indeed to check that the only $\eta$–dependent term which has the quantum numbers of the gauge anomaly would be given by $\int d^4x \, \eta \operatorname{Tr} c^4$, were it not be vanishing due to the cyclicity of the trace and the anticommutativity of the ghost field. Therefore the Slavnov–Taylor identity (6.28) remains unchanged.

For what now concerns the invariant counterterms which one may add to the action (6.45), it turns out that the only possible term depending on the external source $\eta$ has the form (6.44), i.e.,

$$L_{(\eta)} = -i \int d^4x \, \eta \operatorname{Tr} \frac{c^3}{3} = \mathcal{N}_\eta S \, , \tag{6.48}$$

where $\mathcal{N}_\eta$ denotes the counting operator

$$\mathcal{N}_\eta = \int d^4x \, \eta \frac{\delta}{\delta \eta} \, . \tag{6.49}$$

This counterterm is however ruled out by the antighost equation (6.47), implying then the absence of (6.49) in the Callan–Symanzik equation (6.42) which thus remains unchanged, i.e.,

$$\mathcal{C} \Gamma = \hbar^{n+1} \Delta_c^{n+1} + O(\hbar^{n+2}) \, . \tag{6.50}$$

Finally, differentiating the latter with respect to $\eta$ and defining the quantum extension of the composite operator $\operatorname{Tr} c^3$ by

---

[3]The renormalizability of the modified antighost equation (6.47) can be proven using the same procedure as in Subsect. 6.1.1, see also [89].

$$\left[ \left( \mathrm{Tr} \frac{c^3}{3} \right) \cdot \varGamma \right] = i \left. \frac{\delta \varGamma}{\delta \eta} \right|_{\eta=0} \,, \tag{6.51}$$

we get

$$\mathcal{C} \left[ \left( \mathrm{Tr} \frac{c^3}{3} \right) \cdot \varGamma \right] = O(\hbar^{n+1}) \,. \tag{6.52}$$

This means that Green functions with the insertion of $\mathrm{Tr}\, c^3$ obey the same Callan–Symanzik equation as the Green functions without this insertion, i.e., the anomalous dimension of $\mathrm{Tr}\, c^3$ is vanishing. In other words the ghost cocycle $\mathrm{Tr}\, c^3$ is ultraviolet finite. As before, this result extends to all orders in the case of the absence of the gauge anomaly.

Similar arguments [89] can be used to prove the equivalent of (6.52) in the more general case:

$$\mathcal{C} \left[ \left( \mathrm{Tr}\, c^{(2p+1)} \right) \cdot \varGamma \right] = O(\hbar^{n+1}) \,. \tag{6.53}$$

## 6.3 The Adler–Bardeen Theorem

In this section we use the finiteness properties of the ghost cocycles $\mathrm{Tr}\, c^{2p+1}$ in order to present an algebraic proof of the Adler–Bardeen nonrenormalization theorem [118] for the gauge anomaly. This theorem is of fundamental importance for the construction of consistent high energy physics models. It states that the anomaly coefficient $r$ of (6.28) vanishes at all orders of perturbation theory if it vanishes in the one–loop approximation.

Let us recall also (cf. (4.3)) that at the one–loop order the coefficient $r$ is related to a triangle diagram with three gauge fields as external (amputated) legs and turns out to be proportional to

$$d_{(abc)} \mathrm{Tr}\, (T^a T^b T^c) \,, \tag{6.54}$$

where $d_{(abc)}$ is the totally symmetric invariant tensor of rank three and $\{T^a\}$ are the generators of the fermion representation. Therefore if the matter field representation is chosen in such a way that (6.54) vanishes then the nonrenormalization theorem ensures that the anomaly will be definitively absent. This means that the BRS invariance is preserved at the quantum level. This is a necessary requirement in order to have a physical theory with unitary $S$–matrix and gauge invariant observables [15, 16].

### 6.3.1 The Callan–Symanzik Equation for the Anomaly

It has been shown in Sect. 5.5 that the anomaly operator

$$\mathcal{A}(A, c) = \int \Omega_4^1 = \frac{1}{2} \int \mathrm{Tr} \left( dc(AdA + dAA - iA^3) \right) \,, \tag{6.55}$$

as well as the complete set of cocycles $\Omega_p^{5-p}$ of the ladder (5.43) can be renormalized in such a way that they obey the Callan–Symanzik equation (5.77) or (5.79) with a common anomalous dimension $\gamma_u$.

Let us recall indeed that to the cocycles $\Omega_p^{5-p}$ corresponds a set of operator insertions $\Omega_p^{5-p} \cdot \varGamma$ (5.65) which obeys the quantum version (5.66) of the descent equations

$$\mathcal{S}_\Gamma\left[\Omega_p^{5-p}\cdot\Gamma\right]+d\left[\Omega_{p-1}^{4-p}\cdot\Gamma\right]=O(\hbar^n)\,,\qquad 4\geq p\geq 1\,,$$
$$\mathcal{S}_\Gamma[\Omega_0^5\cdot\Gamma]=O(\hbar^n)\,,$$

(6.56)

where $\mathcal{S}_\Gamma$ is the linearized Slavnov–Taylor operator (5.71) (with $\alpha=0$)

$$\mathcal{S}_\Gamma=\int d^4x\left(\ \mathrm{Tr}\,\frac{\delta\Gamma}{\delta\rho^\mu}\frac{\delta}{\delta A_\mu}+\mathrm{Tr}\,\frac{\delta\Gamma}{\delta A_\mu}\frac{\delta}{\delta\rho^\mu}+\mathrm{Tr}\,\frac{\delta\Gamma}{\delta\bar{Y}}\frac{\delta}{\delta\psi}+\mathrm{Tr}\,\frac{\delta\Gamma}{\delta\psi}\frac{\delta}{\delta\bar{Y}}-\mathrm{Tr}\,\frac{\delta\Gamma}{\delta Y}\frac{\delta}{\delta\bar{\psi}}-\mathrm{Tr}\,\frac{\delta\Gamma}{\delta\bar{\psi}}\frac{\delta}{\delta Y}\right.$$
$$\left.+\mathrm{Tr}\,\frac{\delta\Gamma}{\delta\sigma}\frac{\delta}{\delta c}+\mathrm{Tr}\,\frac{\delta\Gamma}{\delta c}\frac{\delta}{\delta\sigma}+\mathrm{Tr}\,B\frac{\delta}{\delta\bar{c}}\right)-\int\sum_{p=0}^{3}du_{3-p}^{p-4}\frac{\delta}{\delta u_{4-p}^{p-5}}\,.$$

(6.57)

According to the procedure of Sect. 5.5, we have introduced an external ladder of forms $u_{4-p}^{p-5}$ $(0\leq p\leq 4)$, of ghost number $p-5$, coupled to the cocycles $\Omega_p^{5-p}$ by adding the term (5.69) to the classical action (6.4).

Notice that the quantum descent equations (6.56) are valid to the same order as the Slavnov–Taylor identity (6.28) since the nilpotency of $\mathcal{S}_\Gamma$ is broken by the gauge anomaly, i.e.,

$$\mathcal{S}_\Gamma\mathcal{S}_\Gamma=O(\hbar^n)\,.$$

(6.58)

Accordingly, for the Callan–Symanzik equation in presence of the fields $u$ we get

$$\mathcal{C}_{(u)}\Gamma=(\mathcal{C}+\hbar\gamma_u\mathcal{N}_u)\,\Gamma\ =\ \hbar\mathcal{S}_\Gamma[\hat{\Delta}\cdot\Gamma]+O(\hbar^{n+1})\,,$$

(6.59)

where $\hat{\Delta}$ is some $u$-dependent insertion of ghost number $-1$, $\mathcal{C}$ is the Callan–Symanzik operator of (6.34) and $\mathcal{N}_u$ is the $u$-counting operator defined in (5.76). The fact that the anomaly starts to produce effects only at the order $\hbar^{n+1}$ follows from the same argument as the one used in Subsect. 6.2.1.

Differentiation of (6.59) with respect to $u_4^{-5}$ and use of the anticommutativity of this derivative with the linearized Slavnov–Taylor operator $\mathcal{S}_\Gamma$, yields the Callan–Symanzik equation for the insertion of the cocycle $\mathrm{Tr}\,c^5$:

$$(\mathcal{C}+\hbar\gamma_u)\left[\left(\mathrm{Tr}\,\frac{c^5}{10}\right)\cdot\Gamma\right]=\hbar\mathcal{S}_\Gamma\left.\left(\frac{\delta}{\delta u_4^{-5}}[\hat{\Delta}\cdot\Gamma]\right)\right|_{u=0}+O(\hbar^{n+1})=O(\hbar^{n+1})\,.$$

(6.60)

The last equality follows from the fact that the most general expression for the $u_4^{-5}$-dependent part of $\hat{\Delta}$ reads $u_4^{-5}\mathrm{Tr}\,c^4$, which identically vanishes. But since the insertion $\mathrm{Tr}\,c^5$ has vanishing anomalous dimension (cf. Subsect. 6.2.2), we conclude that

$$\gamma_u=O(\hbar^n)\,.$$

(6.61)

Differentiating now (6.59) with respect to $u_0^{-1}$ and integrating over space–time we get the Callan–Symanzik equation

$$\mathcal{C}(\mathcal{A}\cdot\Gamma)=-\left.\hbar\mathcal{S}_\Gamma\int\frac{\delta[\hat{\Delta}\cdot\Gamma]}{\delta u_0^{-1}}\right|_{u=0}+O(\hbar^{n+1})\,,$$

(6.62)

for the anomaly insertion

$$\mathcal{A}\cdot\Gamma=\int\Omega_4^1\cdot\Gamma\,,$$

(6.63)

without anomalous dimension, up to an irrelevant $\mathcal{S}_\Gamma$-variation following from the right–hand side of (6.59) and from the anticommutativity of $\int\delta/\delta u_0^{-1}$ with $\mathcal{S}_\Gamma$ which can be read off from the expression (6.57) of the latter operator.

### 6.3.2 The Nonrenormalization Theorem of the Gauge Anomaly

Let us now prove the Adler–Bardeen nonrenormalization theorem [118] for the gauge anomaly.

The first step in the algebraic proof we are going to give is [88] to extend the anomalous Slavnov–Taylor identity (6.28) to the order $\hbar^{n+1}$ as

$$\mathcal{S}(\Gamma) = \hbar^n r(\mathcal{A} \cdot \Gamma) + \hbar^{n+1}\mathcal{D} + O(\hbar^{n+2}) , \tag{6.64}$$

where $\mathcal{A} \cdot \Gamma$ is the renormalized anomaly insertion defined in (6.63) and $\mathcal{D}$ is an integrated local functional of ultraviolet dimension four and ghost number one. The locality of $\mathcal{D}$ follows from the simple observation that, from the Quantum Action Principle, the left hand side of (6.64) defines an insertion and that the difference between two insertions, i.e., $(\mathcal{S}(\Gamma) - \hbar^n r(\mathcal{A} \cdot \Gamma))$, is an insertion and then its lowest order is a local functional.

Applying the Callan–Symanzik operator $\mathcal{C}$ of (6.42) to both sides of (6.64) and using the equation (6.62) for the anomaly insertion $\mathcal{A} \cdot \Gamma$ we get, to the lowest order (i.e., order n+1) in $\hbar$, the condition

$$\left(\beta_g^{(1)}\frac{\partial r}{\partial g}\right)\mathcal{A} = \mu\frac{\partial \mathcal{D}}{\partial \mu} + \mathcal{S}_S\bar{\Delta} , \tag{6.65}$$

where $\mathcal{A}$ is the gauge anomaly expression (4.55), $\beta_g^{(1)}$ is the one–loop gauge beta function and $\bar{\Delta}$ is a local integrated functional.

Taking into account that $\mathcal{D}$ is homogeneous of degree zero in the mass parameters [88][4]

$$\mu\frac{\partial \mathcal{D}}{\partial \mu} = 0 , \tag{6.66}$$

and that the gauge anomaly $\mathcal{A}$ cannot be written as a local $\mathcal{S}_S$–variation, it follows that (6.65) reduces to

$$\beta_g^{(1)}\frac{\partial r}{\partial g} = 0 , \tag{6.67}$$

which, in the generic case of $\beta_g^{(1)} \neq 0$, implies the Adler–Bardeen theorem [120, 121, 87, 88, 89]. Indeed, from the expression of the invariant classical action (4.7) one easily sees that the interaction term between the matter and the gauge fields does not contain the gauge coupling constant $g$ which, on the contrary, explicitly appears in the gauge propagator (as a factor $g^2$) and in the three and quartic gauge self–interaction terms (as a factor $1/g^2$). This implies that the coupling constant behaviour of an $L$–loop one–particle irreducible Feynman diagram is given by a factor $(g^2)^{L-1}$. It is apparent then that to the one–loop approximation the anomalous triangle diagram is independent from the coupling constant $g$. Therefore if with a suitable choice of the matter representation the one–loop triangle diagram is identically zero, the anomaly coefficient $r$ at its lowest order (i.e., $n \geq 2$) will necessarily depend on $g$ and then it will vanish due to (6.67). This proves the nonrenormalization theorem at the order considered and hence to all orders by induction.

---

[4]This property follows from the fact that $\mathcal{D}$ is a polynomial of maximum dimension in the fields, i.e., of dimension 4, which implies that its coefficients are dimensionless.

Let us conclude this chapter by remarking that the Adler–Bardeen theorem also holds in the case in which, besides the gauge coupling $g$, the model contains Yukawa and scalar self—matter couplings [87, 88, 89]. Recently it has been extended to include the case of models with a vanishing one–loop gauge beta function [90], using a generalization of the previous arguments.

*Remark.* The Adler–Bardeen theorem, here proven in the particular case of the Landau gauge, has a general validity and extends to the case of a generic covariant gauge (cf. (4.8)). This follows from the fact that, as already discussed in Subsect. 5.5.2, the anomalous dimension $\gamma_u$ of eq.(6.60) is independent from the gauge parameter(s), this property being shared by the anomaly coefficient $r$ at its lowest order (see Sect. 4.5 and Subsect. 5.2.2).

# 7. Topological Field Theories

Topological field theories are a class of gauge models with the peculiarity that their observables are of topological nature, as for instance the knot and link invariants in the case of three–dimensional Chern–Simons theory [111, 122], the Donaldson invariants for the four–dimensional topological Yang–Mills theory [112], and many other examples [113]. It is worth noticing that all the aforementioned gauge invariant quantities can be computed by the techniques of standard perturbative field theory.

The object of the present chapter is to give an introduction to such computations within the algebraic renormalization approach. We shall see in this way that the topological theories exhibit remarkable ultraviolet finiteness properties. In particular, the models considered here, namely the three–dimensional Chern–Simons and the $BF$ theories, will provide examples of fully finite quantum field theories, i.e., theories with vanishing $\beta$ functions and anomalous dimensions. The latter property relies on the existence of a supersymmetry whose origin is manifest in the case of the Landau gauge. It is generated by a scalar fermionic charge, identified with the BRS operator, and a set of fermionic charges forming a Lorentz vector. Their algebra is of the Wess–Zumino type, i.e., it closes on the space–time translations.

This supersymmetry plays moreover a crucial role in the construction of the explicit solutions of the BRS cohomology modulo $d$, thus giving a systematic classification of all possible anomalies and physically relevant invariant counterterms.

## 7.1 The Chern–Simons Model

The Chern–Simons term in three space–time dimensions is given by the following gauge invariant action

$$S_{\mathrm{CS}} = -\frac{k}{4\pi} \int d^3x \; \varepsilon^{\mu\nu\rho} \mathrm{Tr} \left( A_\mu \partial_\nu A_\rho - \frac{2i}{3} A_\mu A_\nu A_\rho \right) , \tag{7.1}$$

where $k$ denotes the inverse of the coupling constant and $A_\mu$ is a Lie algebra valued gauge field, the corresponding group being chosen to be simple and compact. Expression (7.1), although referred here to the flat Euclidean space–time, has an intrinsic geometrical meaning and can be naturally defined on an arbitrary three dimensional manifold $\mathcal{E}$. Let us remark that the Chern–Simons action (7.1), being the integral of a 3–form, does not depend on the metric $g_{\mu\nu}$ which one can introduce on $\mathcal{E}$. This important feature, as shown by [122], allows to compute topological invariants of three dimensional manifolds by using perturbative field theory techniques. The topological character of the action (7.1) is also at the origin of the ultraviolet finiteness of the perturbative Feynman diagrams expansion. It turns out indeed that expression (7.1) yields a vertex functional $\Gamma$ which obeys a Callan–Symanzik equation with vanishing $\beta$–function and no anomalous dimensions [123, 124]. Therefore the Chern–Simons theory represents an example of a fully finite field theory.

These ultraviolet finiteness properties can be easily understood by the following formal nonperturbative argument. Requiring that the Feynman path integral (2.7) is invariant under the finite gauge transformations (see (4.5))

$$A_\mu^{\mathcal{U}} = \mathcal{U} A_\mu \mathcal{U}^{-1} + i\mathcal{U}\partial_\mu \mathcal{U}^{-1} , \tag{7.2}$$

with

$$\mathcal{U} = e^{i\theta} , \tag{7.3}$$

$\theta$ being a Lie algebra valued local parameter, forces the coupling constant $k$ to be *quantized*, i.e., $k$ can assume only discrete integer values. Indeed, the *quantization* of the coupling constant $k$ follows from the fact that, contrary to the four dimensional Yang–Mills case (4.7), the Chern–Simons action (7.1) is not left invariant by the finite transformations (7.2). Its variation reads

$$S_{\mathrm{CS}}(A^{\mathcal{U}}) - S_{\mathrm{CS}}(A) = -\frac{k}{12\pi}\mathcal{I} , \tag{7.4}$$

where $\mathcal{I}$ is the integral of the second Chern character [125]

$$\mathcal{I} = \int d^3x \; \varepsilon^{\mu\nu\rho}\mathrm{Tr}\left(\mathcal{U}\partial_\mu\mathcal{U}^{-1}\mathcal{U}\partial_\nu\mathcal{U}^{-1}\mathcal{U}\partial_\rho\mathcal{U}^{-1}\right) . \tag{7.5}$$

This quantity, when referred to a nontrivial three dimensional space $\mathcal{E}$, is known to possess a topological meaning, its value being given by

$$\mathcal{I} = 24\pi^2 n , \tag{7.6}$$

where $n$ is an integer associated with the map $\mathcal{U}$ of the three dimensional space $\mathcal{E}$ into the gauge group. Thus $S_{\mathrm{CS}}(A^{\mathcal{U}})$ differs from $S_{\mathrm{CS}}(A)$ by the quantity $(2\pi n k)$. The requirement of invariance of the path integral (2.7) under (7.2) implies then[1]

$$e^{-i2\pi n k} = 1 , \tag{7.7}$$

from which it follows that $k$ is an integer. As a consequence one expects that in perturbation theory $k$ will be not renormalized, i.e., its beta function is expected to vanish.

The above nonperturbative argument cannot be used within a purely perturbative approach, as the one adopted here. As one can easily understand, this is due to the fact that perturbation theory feels only the infinitesimal version of (7.2), i.e.,

$$\mathcal{U} = 1 + i\theta + O(\theta^2) ,$$
$$\delta A_\mu = D_\mu \theta := \partial_\mu \theta + i[\theta, A_\mu] , \tag{7.8}$$

---

[1]Notice that, due to the metric independence of the Chern–Simons action (7.1), the factor $i$ in the path integral measure $e^{iS_{\mathrm{CS}}}$ does not disappear when one moves from the Minkowski to the Euclidean space–time by means of a Wick rotation (cf. Chap. 2).

which, leaving the classical action (7.1) invariant, does not impose any constraint on the possible values of the coupling constant $k$. Moreover, perturbation theory being an expansion at zero coupling constant, is an expansion at infinite $k$.

It turns out however that, once a gauge fixing term of the Landau type has been introduced, the Chern–Simons action possesses a further global invariance, called *vector supersymmetry* [126, 127, 128], whose generators are fermionic and carry a Lorentz index. This new invariance leads together with the BRS invariance to an algebra of the Wess–Zumino type, i.e., to an algebra which closes on the space–time translations. It allows thus for a supersymmetric interpretation of the model. Moreover, when combined with the antighost equation (6.7), it forbids the presence of any invariant local counterterm, yielding then a completely algebraic proof of the ultraviolet finiteness of the model at all orders of perturbation theory [124]. The next sections will explain this in details.

## 7.2 The Vector Supersymmetry

In order to fix the gauge invariance (7.8) of the Chern–Simons action (7.1) we use the Landau gauge (6.1), easily adapted to three dimensional space–time,

$$S_{\text{gf}} = \text{Tr} \int d^3x \ (B\partial^\mu A_\mu - \bar{c}\partial^\mu(\partial_\mu c + i[c, A_\mu])) \ . \tag{7.9}$$

The gauge–fixed action

$$\tilde{S} = S_{\text{CS}} + S_{\text{gf}} \ , \tag{7.10}$$

is invariant under the nilpotent BRS transformations (4.9)

$$sA_\mu = D_\mu c \ , \qquad sc = ic^2 \ ,$$
$$s\bar{c} = B \ , \qquad sB = 0 \ . \tag{7.11}$$

The gauge fixing part of the action of course depends on the metric, chosen here as the flat Euclidean one $\delta_{\mu\nu}$.

*Remark.* One should note that in the present case the use of a general linear covariant $\alpha$–gauge (4.8) is not straightforward. Indeed, contrary to the four dimensional case, the parameter $\alpha$ has now the dimension of a mass. The resulting gauge propagator is given by the sum of a purely transverse term $(\varepsilon_{\mu\nu\rho}p^\rho)/p^2$ (corresponding to $\alpha = 0$) and of a longitudinal part $\alpha p_\mu p_\nu/(p^2)^2$ which, due to the factor $(p^2)^2$ in the denominator, can give rise to infrared divergencies in the Feynman integrals.

As said before the action (7.10), besides the BRS invariance (7.11), possesses an additional global supersymmetry whose generators carry a Lorentz index. In order to derive it let us consider the energy–momentum tensor $T_{\mu\nu}$ of the theory defined by (7.10). The latter, as a consequence of the Chern–Simons action (7.1) being metric independent, is easily seen to be an exact BRS variation

$$T_{\mu\nu} = s\Lambda_{\mu\nu} \ , \tag{7.12}$$

with

$$\Lambda_{\mu\nu} = \text{Tr}\left(-(\partial_\mu \bar{c})A_\nu - (\partial_\nu \bar{c})A_\mu + \delta_{\mu\nu}(\partial^\tau \bar{c})A_\tau\right) . \tag{7.13}$$

Let us compute the divergence of $\Lambda_{\mu\nu}$. Making use of the equation of motion of the gauge field $A$

$$\frac{\delta \tilde{S}}{\delta A_\mu} = -\frac{k}{4\pi}\varepsilon^{\mu\nu\rho}F_{\nu\rho} - \partial^\mu B + i\{\partial^\mu \bar{c}, c\} , \tag{7.14}$$

one gets[2]

$$\partial^\mu \Lambda_{\mu\nu} = \text{Tr}\left(\frac{2\pi}{k}\varepsilon_{\mu\nu\tau}\partial^\mu \bar{c}\frac{\delta \tilde{S}}{\delta A_\tau} - A_\nu \frac{\delta \tilde{S}}{\delta c} - \partial_\nu \bar{c}\frac{\delta \tilde{S}}{\delta B} + \frac{2\pi}{k}\varepsilon_{\mu\nu\tau}\partial^\mu \bar{c}\partial^\tau B\right) . \tag{7.15}$$

This shows that

$$\tilde{\Lambda}_{\mu\nu} = \Lambda_{\mu\nu} - \frac{2\pi}{k}\text{Tr}\left(\varepsilon_{\mu\nu\tau}\bar{c}\partial^\tau B\right) , \tag{7.16}$$

is a conserved quantity since its divergence is equal to contact terms, i.e., to terms which vanish when the equations of motion are used. We also observe that

$$s(\varepsilon_{\mu\nu\tau}\bar{c}\partial^\tau B) = \varepsilon_{\mu\nu\tau}B\partial^\tau B , \tag{7.17}$$

is identically conserved

$$\partial^\mu(\varepsilon_{\mu\nu\tau}B\partial^\tau B) = 0 , \tag{7.18}$$

and can be used as an improvement for the energy–momentum tensor. We can thus redefine the latter:

$$T_{\mu\nu} \to \tilde{T}_{\mu\nu} = T_{\mu\nu} - \text{Tr}\left(\frac{2\pi}{k}\varepsilon_{\mu\nu\tau}B\partial^\tau B\right) , \tag{7.19}$$

so that

$$\tilde{T}_{\mu\nu} = s\tilde{\Lambda}_{\mu\nu} . \tag{7.20}$$

and

$$\partial^\mu \tilde{T}_{\mu\nu} = \text{(contact terms)} , \tag{7.21}$$

as well as

$$\partial^\mu \tilde{\Lambda}_{\mu\nu} = \text{Tr}\left(\frac{2\pi}{k}\varepsilon_{\mu\nu\tau}\partial^\mu \bar{c}\frac{\delta \tilde{S}}{\delta A_\tau} - A_\nu \frac{\delta \tilde{S}}{\delta c} - \partial_\nu \bar{c}\frac{\delta \tilde{S}}{\delta B}\right) . \tag{7.22}$$

Equations (7.19)–(7.22) show that $T_{\mu\nu}$ and $\tilde{T}_{\mu\nu}$ are equivalent and that both $\tilde{T}_{\mu\nu}$ and $\tilde{\Lambda}_{\mu\nu}$ are conserved quantities. This fact is at the origin of the existence of the vector supersymmetry. Indeed, integrating (7.22) on space–time one gets the identity

$$\int d^3x\, \text{Tr}\left(\frac{2\pi}{k}\varepsilon_{\mu\nu\tau}\partial^\mu \bar{c}\frac{\delta \tilde{S}}{\delta A_\tau} - A_\nu \frac{\delta \tilde{S}}{\delta c} - \partial_\nu \bar{c}\frac{\delta \tilde{S}}{\delta B}\right) = 0 , \tag{7.23}$$

which expresses the invariance of the gauge–fixed action (7.10) under the linear transformations

$$\delta_\nu A_\tau = \frac{2\pi}{k}\varepsilon_{\mu\nu\tau}\partial^\mu \bar{c} ,$$

$$\delta_\nu c = -A_\nu ,$$

$$\delta_\nu B = -\partial_\nu \bar{c} , \tag{7.24}$$

$$\delta_\nu \bar{c} = 0 .$$

---

[2]The tensor $\varepsilon^{\mu\nu\rho}$ is normalized as $\varepsilon^{123} = \varepsilon_{123} = 1$. One also has $\varepsilon_{\mu\nu\rho}\varepsilon^{\rho\sigma\tau} = \delta_\mu^\sigma \delta_\nu^\tau - \delta_\nu^\sigma \delta_\mu^\tau$ .

The operator $\delta_\nu$ gives rise together with the BRS operator to the anticommutation relations

$$\{\delta_\mu, \delta_\nu\} = 0 \ ,$$
$$\{\delta_\mu, s\} = -\partial_\mu + (\text{eq. of motion}) \ ,$$ (7.25)

which, closing on–shell on the space–time translations, describe a superalgebra of the Wess–Zumino type. The latter is called a *vector supersymmetry* since the generators $\delta_\mu$ form a Lorentz vector (the fourth generator $s$ is a scalar).

The above construction can be easily repeated for other topological models such as the $BF$ [129, 130] (cf. Sect. 7.4) and the cohomological Witten's models [131, 99]. It turns out that the vector supersymmetry is a general feature of a large class of topological theories. In particular, as we shall see in Chap. 8, it is present also in the bosonic string theory.

### 7.2.1 The Off-Shell Algebra

We now have to write down the Ward identities corresponding to the BRS invariance and to the supersymmetry invariance (7.23). In order to achieve this we introduce, as done in (4.11), external sources $\rho$ and $\sigma$ coupled to the nonlinear BRS variations,

$$S_{\text{ext}} = \int d^3x \ \text{Tr} \left( \rho^\mu D_\mu c + \sigma i c^2 \right) \ ,$$ (7.26)

so that the total classical action

$$S = \tilde{S} + S_{\text{ext}}$$ (7.27)

obeys the Slavnov–Taylor identity

$$\mathcal{S}(S) - \int d^3x \ \text{Tr} \left( \frac{\delta S}{\delta \rho^\mu} \frac{\delta S}{\delta A_\mu} + \frac{\delta S}{\delta \sigma} \frac{\delta S}{\delta c} + B \frac{\delta S}{\delta \bar{c}} \right) = 0 \ .$$ (7.28)

Turning to the supersymmetry, we have to modify the transformations (7.24) because of the source term (7.26). The relevant operator is given by

$$\mathcal{W}_\mu = \int d^3x \ \text{Tr} \left( \frac{2\pi}{k} \varepsilon_{\nu\mu\tau} (\partial^\nu \bar{c} + \rho^\nu) \frac{\delta}{\delta A_\tau} - A_\mu \frac{\delta}{\delta c} - \partial_\mu \bar{c} \frac{\delta}{\delta B} - \sigma \frac{\delta}{\delta \rho^\mu} \right) \ .$$ (7.29)

It yields the linearly broken supersymmetry Ward identity

$$\mathcal{W}_\mu S = \Delta_\mu^{\text{cl}} \ ,$$ (7.30)

with

$$\Delta_\mu^{\text{cl}} = \int d^3x \ \text{Tr} \left( -\frac{2\pi}{k} \varepsilon_{\nu\mu\tau} \rho^\nu \partial^\tau B + \rho^\tau \partial_\mu A_\tau - \sigma \partial_\mu c \right) \ .$$ (7.31)

As in the case of the antighost equation (6.7) the term $\Delta_\mu^{\text{cl}}$, being linear in the quantum fields, is a classical breaking and will not get renormalized.

The on–shell relations (7.25) are replaced now by the following off–shell algebra, valid for any arbitrary functional $\mathcal{F}$ with even ghost number,

$$\mathcal{S}_{\mathcal{F}} \mathcal{S}(\mathcal{F}) = 0 \ ,$$

$$\{\mathcal{W}_\mu, \mathcal{W}_\nu\} = 0 \ , \tag{7.32}$$

$$\mathcal{W}_\mu \mathcal{S}(\mathcal{F}) + \mathcal{S}_{\mathcal{F}}(\mathcal{W}_\mu \mathcal{F} - \Delta_\mu^{\text{cl}}) = -\mathcal{P}_\mu \mathcal{F} \ ,$$

where $\mathcal{S}_{\mathcal{F}}$ and $\mathcal{P}_\mu$ denote respectively the nilpotent linearized Slavnov–Taylor operator (cf. (4.22)) and the translation Ward operator

$$\mathcal{P}_\mu = \sum_{\text{all fields}} \int d^3x \ \text{Tr} \ \partial_\mu \varphi \frac{\delta}{\delta \varphi} \ . \tag{7.33}$$

Moreover, if the functional $\mathcal{F}$ satisfies (7.28) and (7.30) then the algebra of the linearized operators is given by

$$\mathcal{S}_{\mathcal{F}} \mathcal{S}_{\mathcal{F}} = 0 \ , \qquad \{\mathcal{W}_\mu, \mathcal{S}_{\mathcal{F}}\} = -\mathcal{P}_\mu \mathcal{F} \ . \tag{7.34}$$

One thus sees that the introduction of the source term (7.26) closes the algebra (7.25) off–shell[3].

Let us also give, for further use, the algebraic relations involving the Ward operator $\mathcal{W}_\mu$, the Landau gauge condition (6.1), the ghost and antighost equations[4] (4.20), (4.20). They read

$$\mathcal{W}_\mu \left( \frac{\delta \mathcal{F}}{\delta B} - \partial A \right) - \frac{\delta}{\delta B} \left( \mathcal{W}_\mu \mathcal{F} - \Delta_\mu^{\text{cl}} \right) = 0 \ ,$$

$$\mathcal{W}_\mu \mathcal{G} \mathcal{F} + \mathcal{G} \left( \mathcal{W}_\mu \mathcal{F} - \Delta_\mu^{\text{cl}} \right) = 0 \ , \tag{7.35}$$

$$\mathcal{W}_\mu \left( \bar{\mathcal{G}} \mathcal{F} - \Delta_{\text{cl}}^{\mathcal{G}} \right) + \bar{\mathcal{G}} \left( \mathcal{W}_\mu \mathcal{F} - \Delta_\mu^{\text{cl}} \right) = 0 \ .$$

In summary, the complete Chern–Simons action (7.27) obeys:

(1)  the Slavnov–Taylor and the vector supersymmetry Ward identities (7.28), (7.30)

$$\mathcal{S}(S) = 0 \ , \qquad \mathcal{W}_\mu S = \Delta_\mu^{\text{cl}} \ , \tag{7.36}$$

(2)  the Landau gauge condition (6.1) and the ghost equation (4.20)

$$\frac{\delta S}{\delta B} = \partial A \ , \qquad \mathcal{G} S = 0 \ , \tag{7.37}$$

(3)  the antighost equation (6.7) and the rigid gauge invariance (6.15)

$$\bar{\mathcal{G}} S = \Delta_{\text{cl}}^{\mathcal{G}} \ , \qquad \mathcal{R}^{\text{rig}} S = 0 \ , \tag{7.38}$$

with

$$\bar{\mathcal{G}} = \int d^3x \ \left( -i \frac{\delta}{\delta c} + \left[ \bar{c}, \frac{\delta}{\delta B} \right] \right) \ , \qquad \Delta_{\text{cl}}^{\mathcal{G}} = \int d^3x \ ([\rho^\mu, A_\mu] - [\sigma, c]) \ , \tag{7.39}$$

and the rigid Ward operator $\mathcal{R}^{\text{rig}}$ as in (6.14).

Finally, the ghost numbers and the dimensions of all the fields and sources are displayed in Table 7.1.

---

[3]In this sense the external fields $\sigma, \rho$ play the same role as the auxiliary fields of the standard linearly realized supersymmetry [132].

[4]It is easily seen that the antighost equation (6.7) trivially extends to the three dimensional case.

| | $A_\mu$ | $c$ | $\bar{c}$ | $B$ | $\hat{\rho}^\mu$ | $\sigma$ |
|---|---|---|---|---|---|---|
| Ghost number | 0 | 1 | -1 | 0 | -1 | -2 |
| Dimension | 1 | 0 | 1 | 1 | 2 | 3 |

**Table 7.1.** Ghost numbers and dimensions.

## 7.3 Finiteness of the Chern–Simons Theory

Before proving the ultraviolet finiteness of the Chern–Simons model let us begin by showing that the conditions (7.36)––(7.38) which characterize the classical action (7.27) are not anomalous. Rigid invariance (second of eqs. (7.38)) as well as the gauge condition and the ghost equation (7.37) where treated in Chap. 4. The antighost equation (first of eqs. (7.38)) was shown in Chap. 6. The proofs given there applying without modification to the present context we can assume these identities to hold at the quantum level. For the vertex functional $\Gamma$ one has thus

$$\frac{\delta \Gamma}{\delta B} = \partial A , \qquad \mathcal{G}\Gamma = 0 , \tag{7.40}$$

and

$$\bar{\mathcal{G}}\Gamma = \Delta_{\text{cl}}^G , \qquad \mathcal{R}^{\text{rig}}\Gamma = 0 . \tag{7.41}$$

### 7.3.1 Renormalization of the Slavnov–Taylor Identity in Three Dimensions

Let us proceed now with the Slavnov–Taylor identity. As explained in Chap. 4 the quantum extension of the Slavnov–Taylor identity (7.28) is controlled by the cohomology of the corresponding linearized nilpotent operator (4.28)

$$b = \mathcal{S}_S = \int d^3x \; \text{Tr} \left( \frac{\delta S}{\delta \hat{\rho}^\mu} \frac{\delta}{\delta A_\mu} + \frac{\delta S}{\delta A_\mu} \frac{\delta}{\delta \hat{\rho}^\mu} + \frac{\delta S}{\delta \sigma} \frac{\delta}{\delta c} + \frac{\delta S}{\delta c} \frac{\delta}{\delta \sigma} \right) . \tag{7.42}$$

Use has been made of the fact that, due to the gauge condition and the ghost equation (7.40), the operator $b$ acts on the space of integrated local functionals which are independent from the Lagrange multiplier $B$ and depend on the ghost $\bar{c}$ and on the source $\rho_\mu$ only trough the combination $\hat{\rho}_\mu = \rho_\mu + \partial_\mu \bar{c}$ of eq.(4.24).

In order to study the cohomology of $b$ we follow the method of Sect. 5.1. We thus introduce the filtration operator (cf. Subsect. 5.2.1) $\mathcal{N}$

$$\mathcal{N} = \int d^3x \; \text{Tr} \left( A_\mu \frac{\delta}{\delta A_\mu} + \hat{\rho}_\mu \frac{\delta}{\delta \hat{\rho}_\mu} + c \frac{\delta}{\delta c} + \sigma \frac{\delta}{\delta \sigma} \right) , \tag{7.43}$$

according to which $b$ decomposes as

$$b = b_0 + b_1 , \tag{7.44}$$

with

$$b_0 A_\mu = \partial_\mu c \, , \qquad b_0 c = 0 \, ,$$

$$b_0 \hat{\rho}^\mu = -\frac{k}{4\pi} \varepsilon^{\mu\nu\tau} \left( \partial_\nu A_\tau - \partial_\tau A_\nu \right) \, ,$$

$$b_0 \sigma = \partial \hat{\rho} \, ,$$    (7.45)

$$b_0 b_0 = 0 \, .$$

Having in mind to apply Prop. 5.6 of Chap. 5, let us first compute the local cohomology of the operator $b_0$, i.e., the cohomology of $b_0$ in the space of nonintegrated local functionals of the fields and their derivatives. For this purpose we rewrite the transformations (7.45) in a more geometrical way by introducing the duals (see (5.39)) $\hat{\rho}_{\mu\nu}, \sigma_{\mu\nu\tau}$ of the variables $\hat{\rho}_\mu, \sigma$

$$\hat{\rho}_{\mu\nu} = \varepsilon_{\mu\nu\tau} \hat{\rho}^\tau \, , \qquad \sigma_{\mu\nu\tau} = \varepsilon_{\mu\nu\tau} \sigma \, .$$    (7.46)

The equations (7.45) take the form

$$b_0 \sigma_{\mu\nu\tau} = \partial_\mu \hat{\rho}_{\nu\tau} + \partial_\nu \hat{\rho}_{\tau\mu} + \partial_\tau \hat{\rho}_{\mu\nu} \, ,$$

$$b_0 \hat{\rho}_{\mu\nu} = -\frac{k}{2\pi} \left( \partial_\mu A_\nu - \partial_\nu A_\mu \right) \, ,$$    (7.47)

$$b_0 A_\mu = \partial_\mu c \, , \qquad b_0 c = 0 \, .$$

We see that, according to Definition 5.1 and to (5.53), the fields $\sigma_{\mu\nu\tau}, \hat{\rho}_{\mu\nu}, A_\mu, c$ which are forms of degree 3, 2, 1 and 0, respectively, are grouped in a complete ladder[5]. We also notice that the action of the Ward operator $\mathcal{W}_\mu$ on the ladder (7.47) reads

$$\mathcal{W}_\mu c = -A_\mu \, , \qquad \mathcal{W}_\mu A_\nu = \frac{2\pi}{k} \hat{\rho}_{\mu\nu} \, ,$$

$$\mathcal{W}_\mu \hat{\rho}_{\nu\tau} = -\sigma_{\mu\nu\tau} \, , \qquad \mathcal{W}_\mu \sigma_{\nu\tau\rho} = 0 \, ,$$    (7.49)

from which one recognizes that $\mathcal{W}_\mu$ is nothing but the component version of the operator $\nabla$ already introduced in (5.60), (5.61) in order to solve the cohomology of the complete ladder fields. As one can easily understand, properties (7.47), (7.49) rely on the topological character of the Chern–Simons model.

In particular from Proposition 5.13 it follows that the local cohomology of $b_0$ depends only on the ghost field $c$ nondifferentiated, i.e., it is spanned by elements of the type

$$l_{i_1 \ldots i_p} c^{i_1} \ldots c^{i_p} \, ,$$    (7.50)

where the $l_{i_1 \ldots i_p}$ are arbitrary coefficients.

Prop. 5.6 now implies that the local cohomology of the complete linearized Slavnov–Taylor operator $b$ (7.42) is also given by polynomials of the form (7.50). However, invariance under $b$ restricts the coefficients $l_{i_1 \ldots i_p}$ to be invariant tensors of the gauge group. Thus, the local cohomology of $b$ is given by the invariant polynomials generated by the cocycles

---

[5]The first equation in (7.47) is due to the following identity, valid in three dimensions,

$$\varepsilon_{\mu\nu\rho} \partial_\tau - \varepsilon_{\nu\rho\tau} \partial_\mu + \varepsilon_{\rho\tau\mu} \partial_\nu - \varepsilon_{\tau\mu\nu} \partial_\rho = 0 \, .$$    (7.48)

$$\operatorname{Tr} c^{2n+1} , \qquad n \geq 1 . \tag{7.51}$$

We see that, while in the four dimensional Yang–Mills theory the term $\operatorname{Tr} F^2$ (and its generalization (cf. Subsect. 5.3.2)) belong to the local cohomology of the linearized Slavnov–Taylor operator, in the present case, due to the complete ladder structure of the fields and the external sources $\hat{\rho}, \sigma$, the local cohomology of $b$ is independent from the field strength $F$. In addition, the result (7.51) implies the absence of any gauge anomaly in the three dimensional Slavnov–Taylor identity (7.28). This is due to the fact that a possible anomaly cocycle would be related, through the descent equations (5.5), to the local cohomology of $b$ in the sector of form degree zero and ghost number four. However the latter is empty. The absence of gauge anomalies follows then from the vanishing of the local cohomology of $b$ in the sectors with higher form degree $p = (1, 2, 3)$ according to (7.51). Therefore the classical Slavnov–Taylor identity (7.28) can be extended to the quantum level:

$$\mathcal{S}(\Gamma) = 0 . \tag{7.52}$$

It remains to prove the absence of anomalies for the vector supersymmetry.

### 7.3.2 Renormalization of the Vector Supersymmetry

In order to discuss the quantum extension of eq. (7.30) let us write the broken Ward identity

$$\mathcal{W}_\mu \Gamma = \Delta_\mu^{\text{cl}} + \hbar^n \mathcal{A}_\mu \cdot \Gamma ,$$
$$\mathcal{A}_\mu \cdot \Gamma = \mathcal{A}_\mu + O(\hbar) , \tag{7.53}$$

where, according to the Quantum Action Principle, the term $\mathcal{A}_\mu$ – of order $\hbar$ at least – is an integrated local functional with ghost number -1 and dimensions 4. Equations (7.35) imply that $\mathcal{A}_\mu$ is independent from $B$, depends on the fields $\bar{c}$ and $\rho$ only through the combination (4.24), and obeys the homogeneous antighost equation and the rigid gauge invariance

$$\bar{\mathcal{G}} \mathcal{A}_\mu = 0 , \qquad \mathcal{R}^{\text{rig}} \mathcal{A}_\mu = 0 . \tag{7.54}$$

In the same way, from the nonlinear algebra (7.32) one gets the conditions

$$\mathcal{W}_\mu \mathcal{A}_\nu + \mathcal{W}_\nu \mathcal{A}_\mu = 0 , \qquad b \mathcal{A}_\mu = 0 , \tag{7.55}$$

with $b$ the linearized operator (7.42).

It follows from (7.54) that the most general form for $\mathcal{A}_\mu$ with appropriate dimension and ghost number is:

$$\mathcal{A}_\mu = \int d^3x \left( \nu \sigma^a \partial_\mu c^a + \omega^{a(bc)} \hat{\rho}^a_\mu A^b_\nu A^{c\nu} + \lambda^{abc} \hat{\rho}^a_\nu A^{b\nu} A^c_\mu \right.$$
$$\left. + \xi \hat{\rho}^a_\mu \partial A^a + \gamma \hat{\rho}^{a\nu} \partial_\mu A^a_\nu + \tau \hat{\rho}^{a\nu} \partial_\nu A^a_\mu \right) , \tag{7.56}$$

where $\nu, \xi, \gamma, \tau$ are arbitrary parameters and $\omega^{a(bc)}, \lambda^{abc}$ invariant tensors. Using conditions (7.55) one gets

$$\omega^{a(bc)} = 0 , \qquad \lambda^{abc} = -\gamma f^{abc} , \tag{7.57}$$

with $f^{abc}$ the structure constants and

$$\xi = 0 , \qquad \nu = 0 , \qquad \tau = -\gamma . \qquad (7.58)$$

As a consequence the expression for $\mathcal{A}_\mu$ can be written as

$$\mathcal{A}_\mu = \mathcal{W}_\mu \left( \gamma S_{\text{CS}}(A) \right) , \qquad (7.59)$$

where $S_{\text{CS}}$ is the pure Chern–Simons action (7.1).

Equation (7.59) shows that the possible breaking can be reabsorbed by the introduction of a Slavnov–Taylor invariant counterterm which is compatible with conditions (7.40), (7.41). We have thus proven that the vector supersymmetry Ward identity can be extended to the quantum level, i.e.,

$$\mathcal{W}_\mu \Gamma = \Delta_\mu^{\text{cl}} . \qquad (7.60)$$

### 7.3.3 Ultraviolet Finiteness

Let us now look for the arbitrary invariant counterterms which can be be added to the effective action at each order in $\hbar$. The most general one, denoted by $L$, is an integrated local functional with ghost number 0 and dimensions 3 constrained by the stability conditions

$$\frac{\delta L}{\delta B} = 0 , \qquad \mathcal{G}L = 0 , \qquad \mathcal{R}^{\text{rig}}L = 0 , \qquad (7.61)$$

$$bL = 0 , \qquad \mathcal{W}_\mu L = 0 , \qquad (7.62)$$

and

$$\bar{\mathcal{G}}L = 0 . \qquad (7.63)$$

As usual, from (7.61) it follows that $L$ is a rigid invariant functional which is independent from $B$ and depends only on the combination (4.24). Conditions (7.62) are easily worked out, implying that the most general form for $L$ contains only one arbitrary parameter $\zeta$:

$$L = i\zeta \int d^3x \, \text{Tr} \left( \frac{k}{6\pi} \varepsilon^{\mu\nu\rho} A_\mu A_\nu A_\rho + \sigma c^2 + \bar{\rho}^\mu [A_\mu, c] \right) . \qquad (7.64)$$

This expression coinciding with the cubic part of the complete action (7.27), the parameter $\zeta$ represents a possible renormalization of the coupling constant $k$. However, preservation of the homogeneous antighost condition (7.63) gives

$$\zeta = 0 , \qquad (7.65)$$

i.e., there is finally no invariant counterterm compatible with the set of symmetries and constraints (7.36)–(7.38).

This result means that the usual ambiguities due to the renormalization procedure do not appear in the present field theoretical model. This may be interpreted as the ultraviolet finiteness of the topological Chern–Simons theory. A more physical interpretation – in term of scale invariance – will be presented in the next subsection.

*Remarks.*

1. From the result (7.51) on the local cohomology of the operator $b$ it follows that the expression (7.64) is related, through the descent equations (5.5), to the ghost cocycle $\operatorname{Tr} c^3$. Indeed, using the complete ladder structure (7.49) and the general formula (5.59), it is easily checked that the term (7.64) is given by

$$ i \int d^3x \, \epsilon^{\mu\nu\rho} \mathcal{W}_\mu \mathcal{W}_\nu \mathcal{W}_\rho \left( \operatorname{Tr} \frac{c^3}{3!} \right) , \qquad (7.66) $$

which is manifestly supersymmetric invariant. Indeed, the operator $\nabla$ introduced in Sect. 5.4, when written in component form, is nothing else than the supersymmetry operator $\mathcal{W}_\mu$. One verifies moreover that (7.66), although it is a solution of the $b_0$–cohomology, is the solution of the full $b$–cohomology.

2. The cocycle (7.64) can be mapped into the cocycle $\Omega_3^0$ of eq.(5.44) by adding a trivial term. The computation is left as an exercise for the reader.

### 7.3.4 Quantum Scale Invariance

The absence of invariant local counterterms has very strong consequences on the scale dependence of the quantum theory from the renormalization mass $\mu$. Let us try indeed to derive the Callan–Symanzik equation for the effective action $\Gamma$ defined by the Slavnov–Taylor and the supersymmetric Ward identities (7.52), (7.60) and by the conditions (7.40), (7.41).

Commuting (cf. Sect. 4.4) the scale operator $\mu \partial/\partial\mu$ with the above mentioned equations one has

$$ \frac{\delta}{\delta B} \left( \mu \frac{\partial \Gamma}{\partial \mu} \right) = 0 , \qquad \mathcal{G} \left( \mu \frac{\partial \Gamma}{\partial \mu} \right) = 0 , \qquad \mathcal{R}^{\mathrm{rig}} \left( \mu \frac{\partial \Gamma}{\partial \mu} \right) = 0 , \qquad (7.67) $$

$$ \mathcal{S}_\Gamma \left( \mu \frac{\partial \Gamma}{\partial \mu} \right) = 0 , \qquad \mathcal{W}_\mu \left( \mu \frac{\partial \Gamma}{\partial \mu} \right) = 0 , \qquad \bar{\mathcal{G}} \left( \mu \frac{\partial \Gamma}{\partial \mu} \right) = 0 . \qquad (7.68) $$

The use of the quantum action principle (cf. Sect. 3.4.2) implies that the insertion $\Delta \cdot \Gamma$ defined by

$$ \mu \frac{\partial \Gamma}{\partial \mu} = \hbar^p \Delta \cdot \Gamma , \qquad p \geq 1 , $$
$$ \Delta \cdot \Gamma = \Delta + O(\hbar) , \qquad (7.69) $$

at its lowest order $\Delta$ is an integrated local polynomial with ghost number zero and dimension three which, as a consequence of eqs.(7.67), (7.68) has to satisfy the same stability conditions (7.61)–(7.63) as the counterterm $L$. Thus $\Delta = 0$ and

$$ \mu \frac{\partial \Gamma}{\partial \mu} = 0 , \qquad (7.70) $$

i.e., the effective action $\Gamma$ obeys a Callan–Symanzik equation with vanishing $\beta$–function and no anomalous dimensions.

The resulting scale invariance is *the physical interpretation of the "ultraviolet finiteness"*.

*Remark.* Three dimensional power–counting renormalizable matter field actions [133, 134] can be consistently coupled to the pure Chern–Simons term (7.1). In this more general case the $\beta$–function for the Chern–Simons coupling constant $k$ turns out to be still vanishing, the same being however not true for the anomalous dimensions of the fields [135, 136].

## 7.4 *BF* Models

Chern–Simons actions exist in higher (odd) dimensional manifolds [137]. In dimension $D = 2n + 1$, they read (here in the Abelian case)

$$S_{CS(2n+1)} = \int A(dA)^n \, ,$$

and one immediately remarks the absence of quadratic terms for the theories with $n \geq 2$, which makes their quantization very problematic[6], and even unmanageable in the perturbative context of the present book. There is however another class of topological theories of the Schwarz type, namely the $BF$ models, which can be defined in manifolds of any dimension, and which possess a quadratic term in the action.

The classical $BF$ model in a manifold $\mathcal{M}$ of dimension $D = n + 2$, with a gauge group $G$ (chosen here as a simple Lie group) is defined by the classical action

$$S_{BF} = \int_{\mathcal{M}} \mathrm{Tr}\, BF = \frac{1}{2n!} \int d^{n+2}x \,\, \mathrm{Tr}\, \varepsilon^{\mu_1 \cdots \mu_{n+2}} B_{\mu_1 \cdots \mu_n} F_{\mu_{n+1}\mu_{n+2}} \, , \qquad (7.71)$$

where the 2–form[7]

$$F = dA + \frac{i}{2}[A, A] = \frac{1}{2} F_{\mu\nu} dx^\mu dx^\nu$$

is the Yang–Mills field strength of the gauge connection $A = A_\mu dx^\mu$, and $B$ is a $n$–form $B = \frac{1}{n!} B_{\mu_1 \cdots \mu_n} dx^{\mu_1} \cdots dx^{\mu_n}$.

This action being metric independent has a topological meaning [140]. It yields the field equations

$$F = 0 \, , \qquad DB = 0 \, , \qquad (7.72)$$

where $D$ is the covariant external derivative. Its action on a Lie algebra valued $p$–form $\Omega_p$ is defined by

$$D\Omega_p = d\Omega_p + i[A, \Omega_p] \, . \qquad (7.73)$$

The action (7.71) is invariant under two sets of gauge transformations, $\delta_\omega$ and $\delta_\psi$:

---

[6]See however [138] and the recent preprint [139].

[7]The product symbol $\wedge$ is omitted, and the brackets are graded brackets. All fields $\varphi$ in the present section are Lie algebra valued forms. We use the conventions of Chap. 4:

$$\varphi = \varphi^a \tau_a \, , \quad [\tau_a, \tau_b] = i f_{abc} \tau_c \, , \quad \mathrm{Tr}\, \tau_a \tau_b = \delta_{ab} \, .$$

$$\delta_\omega A = D\omega \,, \qquad \delta_\omega B = i[\omega, B] \,,$$
$$\delta_\psi A = 0 \,, \qquad \delta_\psi B = D\psi \,,$$

(7.74)

where $\omega$ and $\psi$ are Lie algebra valued forms of degree 0 and $n-1$, respectively.

*Remark.* In complete analogy with the Chern–Simons theory, the field equations mean that the fields $A$ and $B$ can be gauged away – at least locally. In other words there are no local observables.

Moreover, and contrarily to the Chern–Simons case, there is no coupling constant. Indeed, the action being made of two terms, namely the quadratic and the cubic ones, possesses two parameters – set to the value 1 in (7.71) – which can be both reabsorbed by a multiplicative renormalization of the fields $A$ and $B$.

The gauge fixing and the renormalization of the $BF$ models are studied in the references [115, 141, 142, 129, 130]

### 7.4.1 Gauge Fixing

The Yang–Mills symmetry $\delta_\omega$ can be gauge fixed in the usual way. This is not the case for the symmetry $\delta_\psi$ since it contains zero–modes. Indeed, if $\psi = D\psi'$, where $\psi'$ is an arbitrary $(n-2)$–form, then $\delta_\psi B = DD\psi' = [F, \psi']$ vanishes on–shell due to the equation of motion $F = 0$. Then $\psi' = D\psi''$, with $\psi''$ an arbitrary $(n-3)$–form, is an on–shell zero–mode of the former. This procedure stops when $\psi^{(k-1)} = D\psi^{(k)}$ and $\psi^{(k)}$ is an arbitrary 0–form, *i.e.*, there are $k = n - 1 = d - 3$ such stages. The symmetry $\delta_\psi$ is said to be $k$–stage on–shell reducible.

Fixing a gauge symmetry is more involved when the symmetry is reducible. A possibility would be to follow the Batalin–Vilkovisky prescription [114]. We shall follow a slightly different, but equivalent, procedure. It entails two parts:

I. Turn the parameter $\psi$ of the infinitesimal gauge transformation $\delta_\psi$ into a ghost field by changing its statistics, and then $\psi'$ into a ghost for the ghost $\psi$, and so on. The next step is to define the BRS operator acting on $B$, its ghost and the whole series of ghosts for ghosts. This operator will turn out to be nilpotent, but only on–shell. Finally one introduces external fields – the "antifields" of Batalin–Vilkovisky – coupled to the nonlinear BRS transformations, and one extends the BRS operator on these external fields in a way guarantying its off–shell nilpotency[8].

II. Define the gauge fixing conditions with the help of a suitable set of Lagrange multiplier fields. Then introduce for each of them a corresponding antighost with the usual BRS transformation rules.

Let us begin with part I of the programme. The ghost associated to the gauge invariance $\delta_\omega$ is $\omega$, denoted as usually by $c$, its components $c_a$ being odd Grassmann

---

[8]This procedure is equivalent to solve the master equation of Batalin–Vilkovisky, which will appear below as the Slavnov–Taylor identity (7.82).

numbers. Denoting[9] the $n$–form $B$ by $B_n^0$, its ghost $\psi$ will be denoted by $B_{n-1}^1$, and the ghosts for ghosts by $B_{n-q}^q$ with $2 \geq q \geq n$.

The BRS transformations then read

$$sA = Dc \,, \quad sc = ic^2 \,,$$

$$sB_{n-q}^q = DB_{n-q-1}^{q+1} + i[c, B_{n-q}^q] \,, \quad 0 \leq q \leq n-1 \,, \tag{7.75}$$

$$sB_0^n = i[c, B_0^n] \quad (q = n) \,.$$

This BRS transformation is nilpotent only *on–shell*, since:

$$s^2 B_{n-q}^q = -DDB_{n-q-2}^{q+2} = -i[F, B_{n-q-2}^{q+2}] \,, \quad 0 \leq q \leq n-2 \,, \tag{7.76}$$

and since $F = 0$ is a field equation. Now, as usually (c.f. (4.11) in chap. 4), we introduce the external fields $\gamma \equiv \gamma_{n+1}^{-1}$, $\tau \equiv \tau_{n+2}^{-2}$, $b_{q+2}^{-q-1}$ coupled to the BRS transforms of the fields $A$, $c$, $B_{n-q}^g$, respectively. It is useful to note that the fields and external fields can be arranged in the two arrays (ladders)

$$\phi^{(1)} = \left\{ \phi_p^{(1)}, \ (p = 0, \cdots, n+2) \right\} = \left\{ c, \ A, \ b_{q+2}^{-q-1} \ (q = 0, \cdots, n) \right\} \,,$$
$$\phi^{(2)} = \left\{ \phi_p^{(2)}, \ (p = 0, \cdots, n+2) \right\} = \left\{ B_{n-q}^q \ (q = n, \cdots, 0), \ \gamma, \tau \right\} \,. \tag{7.77}$$

In this notation the *off–shell nilpotent* extension $b$ of the BRS transformations $s$ on the functionals depending on the external fields is then defined by[10]

$$b\phi_p^{(1)} = d\phi_{p-1}^{(1)} + \frac{i}{2} \sum_{q=0}^{n+2} \left[ \phi_q^{(1)}, \phi_{p-q}^{(1)} \right] \,, \quad p = 0, ..., n+2 \,,$$
$$b\phi_p^{(2)} = d\phi_{p-1}^{(2)} + i \sum_{q=0}^{n+2} \left[ \phi_q^{(1)}, \phi_{p-q}^{(2)} \right] \,, \quad p = 0, ..., n+2 \,, \tag{7.78}$$

with

$$b^2 = 0 \,. \tag{7.79}$$

For vanishing external fields, the operator $b$ coincides with $s$ (7.75).

*Remark.* The linear part of the transformations (7.78) has the structure of a ladder as defined in Chap. 5 (Def. 5.1).

Let us now go to part II of the procedure, namely the choice of the gauge fixing conditions. The gauge invariance $\delta_\omega$ needs as usual a Lagrange multiplier $\pi$ and an antighost $\bar{c}$. In order to obtain a complete fixing of the reducible gauge invariance $\delta_\psi$ we have to introduce the set of Lagrange multiplier fields $\Pi_{n-q-k}^{\gamma(k)+1}$ and antighosts $\bar{C}_{n-q-k}^{\gamma(k)}$, for $1 \leq k \leq n$, $0 \leq q \leq n-k$, with $\gamma(k) = q$ for $k$ even and $\gamma(k) = -q-1$ for $k$ odd. They are displayed in Figs. 7.1 and 7.2. The BRS transformations of these fields read

---

[9]In $\varphi_p^q$, the upper index $q$ denotes the ghost number and the lower index $p$ the form degree. The Grassmann parity of $\varphi_p^q$ is defined by the parity of its total degree $q + p$.

[10]By convention $\varphi_p = 0$ if $p < 0$ or $p > n + 2$.

$$
\begin{array}{ccccccc}
 & & & \Pi^0_{n-1} & & & \\
 & & \Pi^1_{n-2} & & \Pi^{-1}_{n-2} & & \\
 & \Pi^0_{n-3} & & \Pi^2_{n-3} & & \Pi^{-2}_{n-3} & \\
\cdots & & \cdots & & \cdots & & \cdots
\end{array}
$$

**Fig. 7.1.** Batalin–Vilkovisky pyramid for the Lagrange multiplier fields. The last line contains the forms of degree 0.

$$
\begin{array}{ccccccccc}
 & & & & B^0_n & & & & \\
 & & & \bar{C}^{-1}_{n-1} & & B^1_{n-1} & & & \\
 & & \bar{C}^0_{n-2} & & \bar{C}^{-2}_{n-2} & & B^2_{n-2} & & \\
 & \bar{C}^{-1}_{n-3} & & \bar{C}^1_{n-3} & & \bar{C}^{-3}_{n-3} & & B^3_{n-3} & \\
\cdots & & \cdots & & \cdots & & \cdots & & \cdots
\end{array}
$$

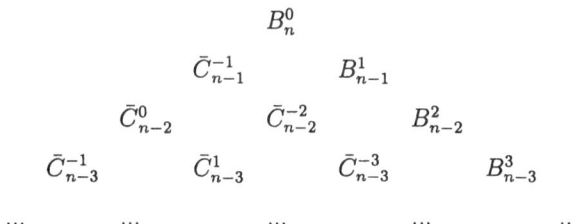

**Fig. 7.2.** Batalin–Vilkovisky pyramid for the field $B$, the ghosts and the antighosts. The last line contains the forms of degree 0.

$$s\bar{c} \equiv b\bar{c} = \pi \ , \qquad\qquad\qquad s\pi \equiv b\pi = 0 \ ,$$

$$s\bar{C}^{\gamma(k)}_{n-q-k} \equiv b\bar{C}^{\gamma(k)}_{n-q-k} = \Pi^{\gamma(k)+1}_{n-q-k} \ , \qquad s\Pi^{\gamma(k)+1}_{n-q-k} \equiv b\Pi^{\gamma(k)+1}_{n-q-k} = 0 \ , \quad 1 \le k \le n \ ,$$

$$0 \le q \le n - k \ . \tag{7.80}$$

Nilpotency is preserved. We then define the gauge fixing by the identities

$$\frac{\delta\Gamma}{\delta\pi} = d * A \ ,$$

$$\frac{\delta\Gamma}{\delta\Pi^{-q}_{n-q-1}} = (-1)^{n+1}d * B^{q}_{n-q} + (-1)^{n+1+q} * d\bar{C}^{q}_{n-q-2}$$

$$+ x_{1,q,n} * \Pi^{q}_{n-q-1} \ , \quad 0 \le q \le n - 1, \ (k = 1) \ ,$$

$$\frac{\delta\Gamma}{\delta\Pi^{\gamma(k)+1}_{n-q-k}} = (-1)^{n+1}d * \bar{C}^{\gamma(k-1)}_{n-q-(k-1)} + (-1)^{n+k+q} * d\bar{C}^{\gamma(k+1)}_{n-q-(k+1)}$$

$$+ \left\{ \begin{array}{ll} (-1)^{q+1}x_{k-1,q+1,n} & (k \text{ even}) \\ x_{k,q,n} & (k \text{ odd}) \end{array} \right\} * \Pi^{\gamma(k-1)}_{n-q-k} \ , \quad 2 \le k \le n,$$

$$0 \le q \le n - k \ , \tag{7.81}$$

where the gauge parameters $x_{k,q,n}$ are (for the moment free) numerical coefficients[11], chosen in such a way that the equations (7.81) are integrable with respect to the variables $\Pi$. One can show that the gauge fixing thus defined is complete, i.e., it leads to well defined free propagators.

## 7.4.2 Slavnov–Taylor Identity and Ghost Equations

BRS invariance of the vertex functional is expressed by the Slavnov–Taylor identity

$$S(\Gamma) := \int_{\mathbb{R}^{n+2}} \text{Tr} \left( \sum_{p=0}^{n+2} \frac{\delta\Gamma}{\delta\phi^{(2)}_{n+2-p}} \frac{\delta\Gamma}{\delta\phi^{(1)}_p} + \pi\frac{\delta\Gamma}{\delta\bar{c}} + \sum_{k=1}^{n}\sum_{g=0}^{n-k} \Pi^{\gamma(k)+1}_{n-g-k} \frac{\delta\Gamma}{\delta\bar{C}^{\gamma(k)}_{n-g-k}} \right) = 0 \ , \tag{7.82}$$

which we have to prove. We have used the ladder notation (7.77). The associated linearized Slavnov–Taylor operator reads

$$\mathcal{S}_\Gamma = \int_{\mathbb{R}^{n+2}} \text{Tr} \left( \sum_{p=0}^{n+2} \left( \frac{\delta\Gamma}{\delta\phi^{(2)}_{n+2-p}} \frac{\delta}{\delta\phi^{(1)}_p} + \frac{\delta\Gamma}{\delta\phi^{(1)}_{n+2-p}} \frac{\delta}{\delta\phi^{(2)}_p} \right) \right.$$

$$\left. + \pi\frac{\delta}{\delta\bar{c}} + \sum_{k=1}^{n}\sum_{g=0}^{n-k} \Pi^{\gamma(k)+1}_{n-g-k} \frac{\delta}{\delta\bar{C}^{\gamma(k)}_{n-g-k}} \right) \ . \tag{7.83}$$

---

[11] The symbol $*$ denotes the Hodge duality, defined for any $p$-form $\omega$ by

$$*\omega = \frac{1}{(d-p)!}(*\omega)_{\beta_1\beta_2\cdots\beta_{d-p}} dx^{\beta_1} dx^{\beta_2} \cdots dx^{\beta_{d-p}} \ ,$$

where

$$(*\omega)_{\beta_1\beta_2\cdots\beta_{d-p}} = \frac{1}{p!}\varepsilon_{\beta_1\beta_2\cdots\beta_{d-p}\alpha_1\cdots\alpha_p}\omega^{\alpha_1\cdots\alpha_p} \ .$$

As usual the validity of the gauge fixing conditions and of the Slavnov–Taylor identity imply a set of ghost equations, which here take the form

$$
\begin{aligned}
&\left(\frac{\delta}{\delta\bar{c}} - d * \frac{\delta}{\delta\gamma}\right)\Gamma = 0 \ , \\
&\left(\frac{\delta}{\delta\bar{C}_{n-g-1}^{-g-1}} - d * \frac{\delta}{\delta b_{g+2}^{-g-1}}\right)\Gamma = (-1)^g * d\Pi_{n-g-2}^{g+1} \ , \qquad 0 \le g \le n-1 \ (k=1) \ , \\
&\frac{\delta}{\delta\bar{C}_{n-g-k}^{\gamma(k)}}\Gamma = d * \Pi_{n-g-(k-1)}^{\gamma(k-1)+1} + (-1)^{k+g+1} * d\Pi_{n-g-(k+1)}^{\gamma(k+1)+1} \ , \qquad 0 \le g \le n-k \ ,
\end{aligned}
$$

$$
2 \le k \le n \ .
$$

(7.84)

Let us first write down the general solution of the gauge conditions (7.81) and of the ghost equations (7.84); we get

$$
\Gamma(\phi^{(1)}, \phi^{(2)}, \bar{c}, \pi, \bar{C}, \Pi) = \hat{\Gamma}(\hat{\phi}^{(1)}, \hat{\phi}^{(2)}) + S_\Pi(A, B, \bar{c}, \pi, \bar{C}, \Pi) \ , \tag{7.85}
$$

where

$$
\begin{aligned}
S_\Pi = \int_{\mathbb{R}^{n+2}} \mathrm{Tr} \Bigg( &\{-d\pi * A - \sum_{g=0}^{n-1} d\Pi_{n-g-1}^{-g} * B_{n-g}^g \\
&+ \sum_{k=2g=0}^{n}\sum^{n-k} \left(-d\Pi_{n-g-k}^{\gamma(k)+1} * \bar{C}_{n-g-k+1}^{\gamma(k-1)} + (-1)^{n+1} d\bar{C}_{n-g-k}^{\gamma(k)} * \Pi_{n-g-k+1}^{\gamma(k-1)+1}\right) \\
&+ \sum_{k=1,3,5,\cdots}^{n}\sum_{g=1}^{n-k} x_{k,g,n}\Pi_{n-g-k}^{-g} * \Pi_{n-g-k}^g \Bigg) \ .
\end{aligned} \tag{7.86}
$$

The truncated action $\hat{\Gamma}$ is the general solution of the homogenous gauge conditions and of the homogenous ghost equations – i.e., of these equations without their right–hand sides. $\hat{\Gamma}$ is hence independent of the multiplier fields $\pi$, $\Pi$ and depends on the antighosts $\bar{c}$, $\bar{C}$ only through the combinations

$$
\begin{aligned}
\hat{\phi}^{(1)} &= \left\{\hat{\phi}_p^{(1)}, \ (p = 0, \cdots, n+2)\right\} = \left\{c, \ A, \ \hat{b}_{q+2}^{-q-1} \ (q = 0, \cdots, n)\right\} \ , \\
\hat{\phi}^{(2)} &= \left\{\hat{\phi}_p^{(2)}, \ (p = 0, \cdots, n+2)\right\} = \left\{B_{n-q}^q \ (q = n, \cdots, 0), \ \hat{\gamma}, \tau\right\} \ ,
\end{aligned} \tag{7.87}
$$

where we have replaced $\gamma$ and $b_q^p$ by the shifted fields

$$
\begin{aligned}
\hat{\gamma} &= \gamma - *d\bar{c} \ , \\
\hat{b}_{g+2}^{-g-1} &= b_{g+2}^{-g-1} + (-1)^{g+1} * d\bar{C}_{n-g-1}^{-g-1} \ , \qquad 0 \le g \le n-1 \ (k=1) \ , \\
\hat{b}_{n+2}^{-n-1} &= b_{n+2}^{-n-1} \ \ (g = n \ , \ k = 1) \ .
\end{aligned} \tag{7.88}
$$

In the tree approximation, the general solution of the Slavnov–Taylor identity is given by the classical gauge fixed action

$$
S(\phi^{(1)}, \phi^{(2)}, \bar{c}, \pi, \bar{C}, \Pi) = \hat{S}(\hat{\phi}^{(1)}, \hat{\phi}^{(2)}) + S_\Pi(A, B, \bar{c}, \pi, \bar{C}, \Pi) \ , \tag{7.89}
$$

where the truncated action $\hat{S}$ is given, up to field renormalizations, by the expression

$$\hat{S} = \int_{\mathbb{R}^{n+2}} \text{Tr} \sum_{p=0}^{n+2} \hat{\phi}_{n+2-p}^{(2)} \left( d\hat{\phi}_{p-1}^{(1)} + \frac{i}{2} \sum_{q=0}^{p} \left[ \hat{\phi}_q^{(1)}, \hat{\phi}_{p-q}^{(1)} \right] \right) \tag{7.90}$$

*Remarks.*

1. The cubic part of the action contains terms quadratic in the external sources, and hence terms quadratic in the antighosts. This is a general features of the theories whose BRS algebra closes only on–shell before the external fields are introduced.

2. One can check that the linearized Slavnov–Taylor operator corresponding to the classical gauge fixed action (7.89) coincides with the BRS operator $b$ defined by (7.78) and (7.80):

$$\mathcal{S}_S = b . \tag{7.91}$$

This identity may be used as a definition of the off–shell nilpotent operator $b$, as we have indeed done for the Chern–Simons theory in the present chapter and for the gauge theories in Chap. 4.

3. A very suggestive notation is the following [115]: one introduces two inhomogeneous forms

$$\mathcal{A} = \sum_{p=0}^{n+2} \hat{\phi}_p^{(1)} , \qquad \mathcal{B} = \sum_{p=0}^{n+2} \hat{\phi}_p^{(2)} , \tag{7.92}$$

and defines the generalized Yang–Mills curvature

$$\mathcal{F} = d\mathcal{A} + \frac{i}{2} [\mathcal{A}, \mathcal{A}] . \tag{7.93}$$

With this the truncated action (7.90) takes the familiar form

$$\hat{S} = \int_{\mathcal{M}} \text{Tr} \ \mathcal{BF}|_{n+2}^0 , \tag{7.94}$$

where $\cdots |_{n+2}^0$ means the restriction to the terms of ghost number 0 and form degree $n+2$.

### 7.4.3 Vector Supersymmetry and Antighost Equation

Like other topological theories, in particular the Chern–Simons theory in the Landau gauge, the $BF$ theory admits a supersymmetry generated by a vector charge [142, 129, 130].

Also here the requirement of the invariance of the action under such a supersymmetry restricts the gauge choice. Indeed, the gauge parameters $x_{k,q,n}$ appearing in the gauge conditions (7.81) and left free up to now, must be given the values

$$x_{k,q,n} = (-1)^{n+q+(k+1)/2} . \tag{7.95}$$

It is only then that a supersymmetry holds. Its action on the (shifted) ladder fields (7.87) turns out to have a very geometrical appearance:

$$\delta^{\mathrm{S}}_{(\xi)}\hat{\phi}^{(A)}_p = -i_\xi \hat{\phi}^{(A)}_{p+1}, \quad p = 0, \cdots, p+2, \quad A = 1, 2, \tag{7.96}$$

where $i_\xi$ is the interior derivative with respect to the (constant) vector field $\xi$. The components $\xi^\mu$ of the latter are the three infinitesimal parameters of this vector supersymmetry.

Supersymmetry acts on the antighosts and multiplier fields as

$$\delta^{\mathrm{S}}_{(\xi)}\bar{c} = 0, \qquad \delta^{\mathrm{S}}_{(\xi)}\pi = \mathcal{L}_\xi \bar{c},$$

$$\delta^{\mathrm{S}}_{(\xi)}\bar{C}^{\gamma(k)}_{n-g-k} = \begin{cases} (-1)^{n+(k+1)/2}g(\xi)\bar{C}^{-g-2}_{n-g-k-1}, & (k \text{ odd}), \\ (-1)^{k/2+1}i_\xi \bar{C}^{g-1}_{n-g-k+1}, & (k \text{ even}), \end{cases}$$

$$\delta^{\mathrm{S}}_{(\xi)}\Pi^{\gamma(k)+1}_{n-g-k} = \begin{cases} (-1)^{n+(k+1)/2}g(\xi)\Pi^{-g-1}_{n-g-k-1} + \mathcal{L}_\xi \bar{C}^{-g-1}_{n-g-k}, & (k \text{ odd}), \\ (-1)^{k/2+1}i_\xi \Pi^g_{n-g-k+1} + \mathcal{L}_\xi \bar{C}^g_{n-g-k}, & (k \text{ even}). \end{cases} \tag{7.97}$$

$\mathcal{L}_\xi = [i_\xi, d]$ is the Lie derivative along $\xi^\mu$. $g(\xi) = g_{\mu\nu}\xi^\mu dx^\nu$ is a 1–form of ghost number 1, the vector field $\xi^\mu$ carrying an odd grading. $g_{\mu\nu}$ is the (flat) space–time metric.

The classical action (7.89) is again not exactly invariant under these supersymmetry transformations. We have indeed

$$\delta^{\mathrm{S}}_{(\xi)}S = \Delta^{\mathrm{S}}_{(\xi)\,\mathrm{cl}}, \tag{7.98}$$

where the breaking

$$\Delta^{\mathrm{S}}_{(\xi)\,\mathrm{cl}} := \int_{\mathbb{R}^{n+2}} \mathrm{Tr}\, \left( -\sum_{g=0}^{n} b^{-g-1}_{g+2}\mathcal{L}_\xi B^g_{n-g} + (-1)^n \gamma \mathcal{L}_\xi A + (-1)^n \tau \mathcal{L}_\xi c \right.$$
$$\left. + d * i_\xi \left( b^{-1}_2 \pi + \gamma \Pi^0_{n-1} \right) \right), \tag{7.99}$$

is classical, i.e., linear in the quantum fields.

At the functional level supersymmetry is expressed by the Ward identities

$$\mathcal{W}^{\mathrm{S}}_{(\xi)}\Gamma = \Delta^{\mathrm{S}}_{(\xi)\,\mathrm{cl}}. \tag{7.100}$$

The gauge fixing was chosen (c.f. (7.95)) by the condition of supersymmetry invariance – more precisely by requiring a supersymmetry Ward identity broken only by terms linear in the quantum fields. It turns out that quite alike the Yang–Mills and Chern–Simons theories in the Landau gauge, the *BF* models with precisely the present gauge choice also obey an antighost equation. But, instead of the coupling of the ghost $c$, this equation characterizes the coupling of the other ghost 0–form, namely $B^n_0$. It reads

$$\bar{\mathcal{G}}\Gamma := \int_{\mathbb{R}^{n+2}} \left( \frac{\delta\Gamma}{\delta B^n_0} + \left[ \frac{\delta\Gamma}{\delta\pi}, \bar{C}^{-n}_0 \right] \right) = \Delta^{\bar{\mathcal{G}}}_{\mathrm{cl}}, \tag{7.101}$$

where

$$\Delta^{\bar{\mathcal{G}}}_{\mathrm{cl}} = \int_{\mathbb{R}^{n+2}} \left( \frac{1}{2}\sum_{p=2}^{n} \left[ b^{-p+1}_p, b^{-n+p-1}_{n+2-p} \right] + \left[ b^{-n}_{n+1}, A \right] + \left[ b^{-n-1}_{n+2}, c \right] \right.$$
$$\left. + \sum_{p=2}^{n}(-1)^{n+p+1} \left[ b^{-p+1}_p, *d\bar{C}^{-n+p-1}_{p-1} \right] \right) \tag{7.102}$$

is again an insertion linear in the quantum fields. Unlike as in the Yang–Mills or in the Chern–Simons case, the application of the antighost equation operator to the Slavnov–Taylor equation does not yield the Ward identity of rigid transformations, but a new Ward identity, broken by linear terms:

$$
\bar{\mathcal{F}}\Gamma := \int_{\mathbb{R}^{n+2}} \left( \sum_{p=0}^{n+2} \left[ \phi_p^{(1)}, \frac{\delta\Gamma}{\delta\phi_p^{(2)}} \right] + \sum_{p=2}^{n} \left[ *d\bar{C}_{n-p+1}^{-p+1}, \frac{\delta\Gamma}{\delta B_p^{n-p}} \right] \right.
$$
$$
\left. + (-1)^{n+1} \left[ \bar{C}_0^{-n}, \frac{\delta\Gamma}{\delta\bar{c}} \right] + (-1)^{n+1} \left[ \Pi_0^{-n+1}, \frac{\delta\Gamma}{\delta\pi} \right] \right) = \Delta_{\mathrm{cl}}^{\bar{\mathcal{F}}} ,
\tag{7.103}
$$

with

$$
\Delta_{\mathrm{cl}}^{\bar{\mathcal{F}}} = \sum_{q=2}^{n} (-1)^{n+p+1} \left[ b_p^{-p+1}, *d\Pi_{p-1}^{-n+p} \right] .
\tag{7.104}
$$

### 7.4.4 Renormalization

In order to control the renormalization of the model, as usual we have first to write down the algebra of the functional operators defined in the preceding section, from which we shall derive the consistency conditions to be imposed on the possible anomalies.

The Slavnov–Taylor operator $\mathcal{S}(F)$ and its linearized form $\mathcal{S}_F$, the Ward identity operators[12] for supersymmetry $\mathcal{W}_{(\xi)}^{\mathrm{S}}$ and for translations[13] $\mathcal{W}_{(\varepsilon)}^{\mathrm{T}}$, as well as the antighost equation operator $\bar{\mathcal{G}}$ and its associated operator $\bar{\mathcal{F}}$ obey the nonlinear algebra

$$
\mathcal{S}_F \mathcal{S}(F) = 0 ,
$$
$$
\mathcal{W}_{(\xi)}^{\mathrm{S}} \mathcal{S}(F) - \mathcal{S}_F \left( \mathcal{W}_{(\xi)}^{\mathrm{S}} F - \Delta_{(\xi)\,\mathrm{cl}}^{\mathrm{S}} \right) = \mathcal{W}_{(\xi)}^{\mathrm{T}} F ,
$$
$$
\mathcal{W}_{(\varepsilon)}^{\mathrm{T}} \mathcal{S}(F) - \mathcal{S}_F \mathcal{W}_{(\varepsilon)}^{\mathrm{T}} F = 0 ,
\tag{7.105}
$$
$$
\bar{\mathcal{G}}\mathcal{S}(F) - (-1)^n \mathcal{S}_F \left( \bar{\mathcal{G}}F - \Delta_{\mathrm{cl}}^{\bar{\mathcal{G}}} \right) = \bar{\mathcal{F}}F - \Delta_{\mathrm{cl}}^{\bar{\mathcal{F}}} ,
$$
$$
\bar{\mathcal{F}}\mathcal{S}(F) + (-1)^n \mathcal{S}_F \left( \bar{\mathcal{F}}F - \Delta_{\mathrm{cl}}^{\bar{\mathcal{F}}} \right) = 0 ,
$$

valid for any functional $F$, and the (anti)–commutation relations[14]

$$
\left[ \mathcal{W}_{(\xi)}^{\mathrm{S}}, \mathcal{W}_{(\xi')}^{\mathrm{S}} \right] = 0 ,
$$
$$
\left[ \bar{\mathcal{G}}, \mathcal{W}_{(\xi)}^{\mathrm{S}} \right] = \left[ \bar{\mathcal{F}}, \mathcal{W}_{(\xi)}^{\mathrm{S}} \right] = 0 ,
\tag{7.106}
$$
$$
\left[ \bar{\mathcal{G}}^a(x), \bar{\mathcal{G}}^b(y) \right] = \left[ \bar{\mathcal{F}}^a(x), \bar{\mathcal{G}}^b(y) \right] = \left[ \bar{\mathcal{F}}^a(x), \bar{\mathcal{F}}^b(y) \right] = 0 .
$$

---

[12]The infinitesimal supersymmetry parameters $\xi^\mu$ are anticommuting numbers, the translation parameters $\varepsilon^\mu$ are commuting.

[13]Translation invariance is obvious. It is expressed by the Ward identity

$$
\mathcal{W}_{(\varepsilon)}^{\mathrm{T}} := \int_{\mathbb{R}^{n+2}} \mathrm{Tr} \sum_{\text{all fields } \varphi} \mathcal{L}_\varepsilon \varphi \frac{\delta}{\delta\varphi} ,
$$

where $\mathcal{L}_\varepsilon = [i_\varepsilon, d]$ is the Lie derivative along the vector field $\varepsilon$. For constant $\varepsilon$: $\mathcal{L}_\varepsilon \varphi = \varepsilon^\mu \partial_\mu \varphi$.

[14]Trivial commutation relations involving the translation operator are not written.

Furthermore if the functional $F$ is the classical action $S$, solution of the Slavnov–Taylor identity and of all the other Ward identities, then to (7.105) corresponds the linear algebra

$$b^2 = 0 \ ,$$

$$\left[ \mathcal{W}^{\text{S}}_{(\xi)}, b \right] = \mathcal{W}^{\text{T}}_{(\xi)} \ , \qquad \left[ \mathcal{W}^{\text{T}}_{(\varepsilon)}, b \right] = 0 \ , \qquad (7.107)$$

$$\left[ \bar{\mathcal{G}}, b \right] = \bar{\mathcal{F}} \ , \qquad \left[ \bar{\mathcal{F}}, b \right] = 0 \ ,$$

with $b$ defined by (7.91).

As we know from our previous experience a convenient way to derive the consistency conditions on the possible anomalies and to solve the symmetry constraints obeyed by the counterterms, is to introduce a generalized nilpotent BRS operator summarizing all the constraints. In the present case it reads

$$\delta = b + \mathcal{W}^{\text{S}}_{(\xi)} + \mathcal{W}^{\text{T}}_{(\varepsilon)} + \text{Tr} \left( u \bar{\mathcal{G}} + v \bar{\mathcal{F}} \right) + \xi^\mu \frac{\partial}{\partial \varepsilon^\mu} + \text{Tr} \left( u \frac{\partial}{\partial v} \right) \ , \qquad (7.108)$$

where the "global ghosts" $\xi^\mu$, $\varepsilon^\mu$, $u^a$ and $v^a$ are the infinitesimal parameters of the supersymmetry transformations, of the translations, of the ghost equation and of its associated equation. Their ghost numbers are 2, 1, $n$ and $n-1$ , respectively, so that $\delta$ has ghost number one. Their grading is equal to the parity of their ghost numbers[15]. The transformation laws of the global ghosts are expressed by the last two terms in (7.108). This together with the algebra (7.106), (7.107) makes $\delta$ a coboundary operator:

$$\delta^2 = 0 \ . \qquad (7.109)$$

The problem of the anomalies amounts to compute the cohomology of $\delta$ in the sector of ghost number 1. The sector of ghost number 0 corresponds to the determination of the counterterms. Let us denote by

$$\Delta^q = \int_{\mathbb{R}^{n+2}} \Omega^q_{n+2} \ , \qquad q = 1, 0 \ , \qquad (7.110)$$

the respective cocycles.

The space on which the coboundary operator acts is here the space of integrated local field polynomials, depending on the global ghosts, too.

The problem of the $\delta$–cohomology will be solved using Prop. 5.6 of Chap. 5. We define the filtration operator

$$\mathcal{N} = \sum_{\text{all fields } \varphi} \int_{\mathbb{R}^{n+2}} \varphi \frac{\delta}{\delta \varphi} + \sum_{g = \varepsilon^\mu, \xi^\mu, v, u} 2g \frac{\partial}{\partial g}. \qquad (7.111)$$

The component $\delta_0$ of the coboundary operator $\delta$, i.e., the part of $\delta$ which commutes with $\mathcal{N}$ (see (5.12)), reads

$$\delta_0 = \int_{\mathbb{R}^{n+2}} \left( \sum_{A=1}^{2} \sum_{p=1}^{n+2} d\phi^{(A)}_{p-1} \frac{\delta}{\delta \phi^{(A)}_p} + \pi \frac{\delta}{\delta \bar{c}} + \sum_{k=1}^{n} \sum_{q=0}^{n-k} \Pi^{\gamma(k)+1}_{\nu-q-k} \frac{\delta}{\delta \bar{C}^{\gamma(k)}_{n-q-k}} \right)$$
$$+ \xi^\mu \frac{\partial}{\partial \varepsilon^\mu} + \text{Tr} \left( u \frac{\partial}{\partial v} \right) \ . \qquad (7.112)$$

---

[15]Now: $\xi$ is commuting and $\varepsilon$ is anticommuting

Remarking that the antighosts and the Lagrange multiplier fields form $\delta_0$–doublets according to Def. 5.7 of Chap. 5, we can discard them from the discussion, owing to Prop. 5.8 which states that the cohomology does not depend on such doublets. The same conclusion holds for the pairs of global ghosts $(\varepsilon^\mu, \xi^\mu)$ and $(v, u)$.

What remains are the two ladders of fields (7.77), whose $\delta_0$–variations are given by the linear part of the transformations (7.78). At the level of the integrands $\Omega_{n+2}^q$ of (7.110) we have to solve the cohomology of $\delta_0$ modulo $d$. We can then apply Prop. 5.14 of Chap. 5. It follows that this cohomology is given by

$$\Omega_{n+2}^q = \frac{1}{(n+2)!} \nabla^{n+2} \Omega_0^{q+n+2} , \tag{7.113}$$

where the 0–forms $\Omega_0^{q+n+2}$ belong to the cohomology of $\delta_0$. We know from Prop. 5.13 that they are polynomials in the 0–form components $\phi_0^{(A)}$ of the ladders, i.e., of the ghost fields $c$ and $B_0^n$, undifferentiated. $\nabla$ is the operator defined in (5.60); however, in order to guaranty that the final results obeys the ghost equations (7.84), we shall express it in terms of the shifted ladder fields (7.87). We may then write $\nabla$ as

$$\nabla \hat{\phi}_p^{(A)} = (p+1)\hat{\phi}_{p+1}^{(A)} = -\delta_{(\xi)}^S \hat{\phi}_p^{(A)}\Big|_{\xi^\mu = dx^\mu} . \tag{7.114}$$

The second equality results from the definition (7.96) of the supersymmetry transformations, where one has formally substituted the differentials $dx^\mu$ to the components of the vector field $\xi$.

Let us now go to the cohomology – in the space of the integrated local functionals – of the full coboundary operator $\delta$, which is isomorphic to a subspace of the cohomology of $\delta_0$. We are going to show that representatives of this cohomology are given by cocycles $\Delta^q$ of the form (7.110), each of their integrands $\Omega_{n+2}^q$ being given by the formula (7.113) with $\Omega_0^{q+n+2}$ a polynomial in $c$ and $B_0^n$ to be specified.

$\Delta^q$ has to be invariant under $\delta$. Since $\Delta^q$ does not depend on the global ghosts, this is equivalent to its separate invariance under the action of the operators $b$, $\mathcal{W}_{(\xi)}^S$, $\mathcal{W}_{(\varepsilon)}^T$, $\bar{\mathcal{G}}$ and $\bar{\mathcal{F}}$.

Invariance under $\mathcal{W}_{(\varepsilon)}^T$ (translations) is straightforward.

Invariance $\bar{\mathcal{G}}$ is equivalent to the condition (see (7.101))

$$\int_{\mathbb{R}^{n+2}} \frac{\delta \Delta^q}{\delta B_0^n} = 0 , \tag{7.115}$$

since there is no dependence on the Lagrange multiplier fields. This implies that $\Omega_0^{q+n+2}$ is a polynomial in $c$ only:

$$\Omega_0^{q+n+2} = t_{a_1 \cdots a_{q+n+2}} c^{a_1} \cdots c^{a_{q+n+2}} . \tag{7.116}$$

Invariance under supersymmetry follows immediately from (7.114), which shows that the operator $\nabla$ is nothing but the action of a supersymmetry generator, and from the fact that the action of $n+3$ supersymmetry generators gives zero since they are anticommuting and only $n+2$ of them are independent.

The invariance under $b$ is the consequence of the commutation relation

$$[b, \nabla] = d , \tag{7.117}$$

provided that the 0–form $\Omega_0^{q+n+2}$ is $b$–invariant. The latter condition implies that the coefficients $t_{a_1 \cdots a_{q+n+2}}$ in (7.116) must be the components of an invariant tensor of the gauge group.

Invariance under $\bar{\mathcal{F}}$ follows immediately from the former invariances due to the commutation rule $[\bar{\mathcal{G}}, b] = \bar{\mathcal{F}}$ (see (7.107)).

This terminates the general discussion of the anomalies and of the nontrivial invariant counterterms.

The actual presence of invariant tensors $t_{a_1 \cdots a_{q+n+2}}$ depends on the space–time dimension $D = n + 2$, on the value of the ghost number $q$ (0 or 1) and on the gauge group. Let us give some examples.

1.  $D = 2 \ (n = 0)$:  $\Delta^0 = 0, \ \Delta^1 = \int_{\mathbb{R}^2} \nabla^2 \mathrm{Tr}\, c^3$.
    It turns out in fact that in this 2–dimensional case the anomaly cocycle $\Delta^1$ is forbidden by the existence of further symmetries [143].

2.  $D = 3 \ (n = 1)$:  $\Delta^0 = \int_{\mathbb{R}^3} \nabla^3 \mathrm{Tr}\, c^3, \ \Delta^1 = 0$.
    Here the counterterm $\Delta^0$ is eliminated by a second antighost equation [129].

3.  $D = 4 \ (n = 2)$:  $\Delta^0 = 0, \ \Delta^1 = \int_{\mathbb{R}^4} \nabla^4 \mathrm{Tr}\, c^5$.
    The cohomology in the anomaly sector is not empty. However $\mathrm{Tr}\, c^5$ involving the symmetric invariant tensor $d_{(abc)}$ cannot appear since all fields belong to the adjoint representation with the consequence that only the antisymmetric tensor $f_{[abc]}$ is involved in the Feynman rules. There is thus no anomaly. And since there is no nontrivial counterterm, this theory is finite.

4.  $D = 5 \ (n = 3)$:  $\Delta^0 = \int_{\mathbb{R}^5} \nabla^5 \mathrm{Tr}\, c^5, \ \Delta^1 = 0$.
    The same conclusion as in the four dimensional case holds – the roles of the anomaly and of the nontrivial counterterm being exchanged.

In higher dimensions candidates for anomalies or nontrivial counterterms begin to appear. That their coefficients turn out to vanish or not is not known. E.g. an invariant $\mathrm{Tr}\, c^7$ involving only the tensor $f_{[abc]}$ exists. It can give rise either to an anomaly of the 6-dimensional theory or to a counterterm of the 7-dimensional theory.

# 8. The Bosonic String

In this chapter we discuss, as a last example of the algebraic renormalization technique, the quantization of the bosonic string theory considered as a gauge theory for the gauge group $(Diff \times Weyl)$, $Diff$ and $Weyl$ denoting respectively the diffeomorphism and the Weyl scale transformations.

Needless to say string theory *(and its supersymmetric version)* has been the source of continuous investigations for several decades and has originated a large number of fruitful applications such as the recent developments in conformal field theories and in lower–dimensional gravity.[1]

We shall limit the discussion here to a few aspects of quantization of the string theory. We shall focus in particular on the BRS treatment of the bosonic string based on the so–called Beltrami parametrization of the two–dimensional world sheet metric $g_{\alpha\beta}$.

This parametrization, introduced by [145, 146, 147], turns out to be quite useful for describing the string properties within the perturbative field theoretical framework. It allows one to use a quantization procedure completely analogous to that of the Yang–Mills theories (cf. Chap 4).

Moreover, as shown in [146], the Beltrami parametrization is the most natural parametrization which exhibits the holomorphic factorization of the Green functions with insertion of the energy–momentum tensor, according to the Belavin–Polyakov–Zamolodchikov scheme [148].

Indeed the Beltrami parameter can be interpreted as the classical source for the $(T_{zz}, T_{\bar{z}\bar{z}})$–components of the energy–momentum tensor. It turns out then that the Slavnov–Taylor identity corresponding to the BRS invariance of the theory can be taken as the starting point for an algebraic characterization of the energy–momentum current algebra.

Let us also mention that, as discussed in detail in [145], the use of the Beltrami parameter allows to eliminate from the very beginning the degree of freedom associated with the Weyl symmetry, the remaining diffeomorphism transformations being kept as the basic local invariance of the string action.

However it is well known that diffeomorphism invariance cannot be preserved at the quantum level. There is indeed an anomaly whose numerical coefficient is, in general, non zero. Its presence would imply the existence of unphysical negative norm states in the Fock space of the string excitations [149]. In order to have a consistent theory one is forced then to impose the vanishing of this coefficient. This requirement

---

[1]For a general exposition on string theory we remind the reader of the excellent books and reviews available in the literature; see for instance [144].

fixes uniquely the critical dimension of the bosonic string to be equal to the famous value of 26 [150].

## 8.1 The Bosonic String in the Beltrami Parametrization

Let us start by considering the bosonic string action

$$S_{\text{inv}} = \frac{1}{2} \int_{\mathcal{M}} d^2 x \sqrt{g} g^{\alpha\beta} \partial_\alpha \mathbf{X} \cdot \partial_\beta \mathbf{X} , \tag{8.1}$$

where $g_{\alpha\beta}$ ($\alpha$, $\beta = 1, 2$) is a metric on the two–dimensional string world sheet $\mathcal{M}$ and $\mathbf{X} = \{X^i, i = 1, \cdots D\}$ are the string coordinates which map $\mathcal{M}$ into the $D$–dimensional flat space $\mathbb{R}^D$.
Introducing a system of complex coordinates $(z, \bar{z})$

$$z = x + iy , \qquad \bar{z} = x - iy , \tag{8.2}$$

the world sheet metric $g$ can be expressed as [145, 146, 147]

$$ds^2 = g_{11} dx^2 + 2 g_{12} dx dy + g_{22} dy^2 = \rho^2 |\, dz + \mu d\bar{z} \,|^2 , \tag{8.3}$$

where, after an easy computation,

$$\rho^2 = \frac{1}{4}(g_{11} + g_{22} + 2\sqrt{g}) , \qquad g = \det(g_{\alpha\beta}) , \tag{8.4}$$

and

$$\mu = \frac{g_{11} - g_{22} + 2ig_{12}}{g_{11} + g_{22} + 2\sqrt{g}} . \tag{8.5}$$

The parameter $\mu$ in (8.3), (8.5) is called a Beltrami differential and is associated with a one–form vector–valued geometrical object $\hat{\mu}$ [151]:

$$\hat{\mu} = \mu d\bar{z} \otimes \frac{\partial}{\partial z} , \qquad \mu = \mu^z_{\ \bar{z}} . \tag{8.6}$$

Denoting with $(\partial, \bar{\partial})$ the partial derivatives with respect to $(z, \bar{z})$,

$$\partial = \partial_z \qquad , \qquad \bar{\partial} = \partial_{\bar{z}} , \tag{8.7}$$

the action (8.1) becomes

$$S_{\text{inv}} = \int_{\mathcal{M}} dz d\bar{z} (1 - \mu\bar{\mu}) \mathcal{D}\mathbf{X} \cdot \overline{\mathcal{D}}\mathbf{X} , \tag{8.8}$$

where $(\mathcal{D}, \overline{\mathcal{D}})$ are the *covariant derivatives*

$$\mathcal{D} = \frac{1}{1 - \mu\bar{\mu}}(\partial - \bar{\mu}\bar{\partial}) , \qquad \overline{\mathcal{D}} = \frac{1}{1 - \mu\bar{\mu}}(\bar{\partial} - \mu\partial) . \tag{8.9}$$

One has to observe that the scaling factor $\rho$ of (8.4) has been eliminated from the new form (8.8) of the action. Indeed, due to the Weyl invariance ($g(x) \rightarrow e^{\varphi(x)} g(x)$),

$S_{\text{inv}}$ depends only on a conformal class of metrics $\{g\}$. The latter, as one can see from (8.3), is naturally parametrized by the Beltrami differential $\mu$.

By construction, the classical action (8.8) is invariant under the infinitesimal diffeomorphism transformations generated by a two–components vector field $(\gamma^z, \gamma^{\bar{z}})$. These transformations, when written in terms of $\mu$, take the form [147, 152]:

$$\delta \mathbf{X} = \gamma^z \partial \mathbf{X} + \gamma^{\bar{z}} \bar{\partial} \mathbf{X} \ ,$$

$$\delta \mu = \gamma^z \partial \mu + \gamma^{\bar{z}} \bar{\partial} \mu + \mu \bar{\partial} \gamma^z + \bar{\partial} \gamma^z - \mu \partial \gamma^z - \mu^2 \partial \gamma^{\bar{z}} \ , \tag{8.10}$$

and

$$\delta S_{\text{inv}} = 0 \ . \tag{8.11}$$

In order to gauge–fix the diffeomorphism invariance we follow the general BRS procedure and, as a first step, we associate to (8.10) a BRS operator $(\delta \to s)$ by turning the local parameters $(\gamma^z, \gamma^{\bar{z}})$ into anticommuting ghost variables transforming as

$$\begin{aligned} s\gamma^z &= \gamma^z \partial \gamma^z + \gamma^{\bar{z}} \bar{\partial} \gamma^z \ , \\ s\gamma^{\bar{z}} &= \gamma^z \partial \gamma^{\bar{z}} + \gamma^{\bar{z}} \bar{\partial} \gamma^{\bar{z}} \ , \end{aligned} \tag{8.12}$$

so that

$$s^2 = 0 \ . \tag{8.13}$$

Before introducing the Faddeev–Popov gauge–fixing term it is convenient to switch from $(\gamma^z, \gamma^{\bar{z}})$ to another set of more convenient variables $(c^z, \bar{c}^{\bar{z}})$ [146]:

$$c^z = \gamma^z + \mu \gamma^{\bar{z}} \ , \qquad \bar{c}^{\bar{z}} = \gamma^{\bar{z}} + \bar{\mu} \gamma^z \ . \tag{8.14}$$

With this choice the BRS transformations (8.10), (8.12) take the simple remarkable form:

$$s\mathbf{X} = \frac{c^z - \mu \bar{c}^{\bar{z}}}{1 - \mu \bar{\mu}} \partial \mathbf{X} + \frac{\bar{c}^{\bar{z}} - \bar{\mu} c^z}{1 - \mu \bar{\mu}} \bar{\partial} \mathbf{X} \ , \tag{8.15}$$

and

$$\begin{aligned} sc^z &= c^z \partial c^z \ , \\ s\bar{c}^{\bar{z}} &= \bar{c}^{\bar{z}} \bar{\partial} \bar{c}^{\bar{z}} \ , \\ s\mu &= \bar{\partial} c^z + c^z \partial \mu - \mu \partial c^z \ , \\ s\bar{\mu} &= \partial \bar{c}^{\bar{z}} + \bar{c}^{\bar{z}} \bar{\partial} \bar{\mu} - \bar{\mu} \bar{\partial} \bar{c}^{\bar{z}} \ . \end{aligned} \tag{8.16}$$

One sees in (8.16) that the introduction of the new ghost variables (8.14) allows to split the fields $(\mu, \bar{\mu}, c, \bar{c})$ into two pairs $(\mu, c)$ and $(\bar{\mu}, \bar{c})$ which do not get mixed under the action of the BRS operator. This important property has been proven to be at the origin of the holomorphic factorization of the string Green's functions [146]. Then, from now on, we shall take the combinations $(c, \bar{c})$ as fundamental variables.

In order to fix the gauge we introduce a pair of antighosts $(b_{zz}, \bar{b}_{\bar{z}\bar{z}})$ and of Lagrange multipliers $(\chi_{zz}, \bar{\chi}_{\bar{z}\bar{z}})$, transforming under BRS as

$$\begin{aligned} sb_{zz} &= \chi_{zz} \ , && s\chi_{zz} = 0 \ , \\ s\bar{b}_{\bar{z}\bar{z}} &= \bar{\chi}_{\bar{z}\bar{z}} \ , && s\bar{\chi}_{\bar{z}\bar{z}} = 0 \ , \end{aligned} \tag{8.17}$$

and we choose the gauge condition [147]

$$\mu = \mu_{\text{cl}} , \tag{8.18}$$

where $\mu_{\text{cl}}$ is a classical prescribed Beltrami differential left invariant by the BRS operator

$$s\mu_{\text{cl}} = 0 . \tag{8.19}$$

In a Landau type gauge, the gauge–fixing action $S_{\text{gf}}$ then reads

$$S_{\text{gf}} = -\int dz d\bar{z} \; s \left( b_{zz}(\mu - \mu_{\text{cl}}) + \bar{b}_{\bar{z}\bar{z}}(\bar{\mu} - \bar{\mu}_{\text{cl}}) \right) , \tag{8.20}$$

i.e.,:

$$S_{\text{gf}} = -\int dz d\bar{z} \left( \chi_{zz}(\mu - \mu_{\text{cl}}) + \bar{\chi}_{\bar{z}\bar{z}}(\bar{\mu} - \bar{\mu}_{\text{cl}}) \right)$$
$$+ \int dz d\bar{z} \left( b_{zz}(\bar{\partial}c^z + c^z \partial\mu - \mu\partial c^z) + \bar{b}_{\bar{z}\bar{z}}(\partial\bar{c}^{\bar{z}} + \bar{c}^{\bar{z}}\bar{\partial}\bar{\mu} - \bar{\mu}\bar{\partial}\bar{c}^{\bar{z}}) \right) , \tag{8.21}$$

and

$$s \left( S_{\text{inv}} + S_{\text{gf}} \right) = 0 . \tag{8.22}$$

*Remark.* It is worth to recall here that the condition (8.18) completely fixes the components of the metric appearing in the invariant action (8.8). As it is well known [144], the world sheet metric $g_{\alpha\beta}$ can be completely gauge–fixed due to the fact that, in two dimensions, the number of independent components of $g$, i.e., $(g_{11}, g_{22}, g_{12})$ or equivalently $(\rho, \mu, \bar{\mu})$, equals the number of local invariances of the string action, namely the diffeomorphism and the Weyl transformations. Diffeomorphism invariance (8.10), being parametrized by a two–component vector field $(\gamma^z, \gamma^{\bar{z}})$, allows indeed to fix two components of $g_{\alpha\beta}$, the remaining one being eliminated by the Weyl invariance $(g \to e^{\varphi(x)}g)$.

As usual, coupling the nonlinear BRS variations of $\{\mathbf{X}\}$ and $(c, \bar{c})$ to invariant external sources $(\mathbf{Y}, L, \bar{L})$

$$S_{\text{ext}} = \int dz d\bar{z} \left( \mathbf{Y}_{z\bar{z}} \cdot s\mathbf{X} + L_{zz\bar{z}}c^z \partial c^z + \bar{L}_{\bar{z}\bar{z}z}\bar{c}^{\bar{z}}\bar{\partial}\bar{c}^{\bar{z}} \right) , \tag{8.23}$$

and using the algebraic property

$$s\mu = \frac{\delta(S_{\text{inv}} + S_{\text{gf}})}{\delta b_{zz}} , \tag{8.24}$$

one gets (suppressing for simplicity all the $(z, \bar{z})$–indices) the classical Slavnov–Taylor identity

$$\int dz d\bar{z} \left( \frac{\delta \overline{S}}{\delta \mathbf{Y}} \cdot \frac{\delta \overline{S}}{\delta \mathbf{X}} + \frac{\delta \overline{S}}{\delta \mu}\frac{\delta \overline{S}}{\delta b} + \frac{\delta \overline{S}}{\delta \bar{\mu}}\frac{\delta \overline{S}}{\delta \bar{b}} + \frac{\delta \overline{S}}{\delta L}\frac{\delta \overline{S}}{\delta c} + \frac{\delta \overline{S}}{\delta \bar{L}}\frac{\delta \overline{S}}{\delta \bar{c}} + b\frac{\delta \overline{S}}{\delta \chi} + \bar{b}\frac{\delta \overline{S}}{\delta \bar{\chi}} \right) = 0 , \tag{8.25}$$

where $\overline{S}$ denotes the complete action

$$\overline{S} = S_{\text{inv}} + S_{\text{gf}} + S_{\text{ext}} . \tag{8.26}$$

The Slavnov–Taylor identity (8.25) can be cast in a more convenient form if one eliminates the two Lagrange multipliers. Indeed, since $(\chi, \bar{\chi})$ enter only linearly and without derivative in the complete action $\bar{S}$, it follows that they can be completely eliminated by using the corresponding equation of motion $\mu = \mu_{\text{cl}}$. This amounts to replace everywhere the original Beltrami parameter $\mu$ with the classical one $\mu_{\text{cl}}$.

The Slavnov–Taylor identity (8.25) takes now the simpler form

$$\mathcal{S}(S) = 0 \ , \tag{8.27}$$

with

$$\mathcal{S}(S) = \int dz d\bar{z} \left( \frac{\delta S}{\delta \mathbf{Y}} \cdot \frac{\delta S}{\delta \mathbf{X}} + \frac{\delta S}{\delta \mu_{\text{cl}}} \frac{\delta S}{\delta b} + \frac{\delta S}{\delta L} \frac{\delta S}{\delta c} + \frac{\delta S}{\delta \bar{\mu}_{\text{cl}}} \frac{\delta S}{\delta \bar{b}} + \frac{\delta S}{\delta \bar{L}} \frac{\delta S}{\delta \bar{c}} \right) \ . \tag{8.28}$$

and $S$ the complete action (8.26) computed at $\mu = \mu_{\text{cl}}$:

$$S = \bar{S}|_{\mu=\mu_{\text{cl}}} \ , \tag{8.29}$$

i.e.,:

$$S = \left( S_{\text{inv}} + S_{\text{ext}} \right)\Big|_{\mu=\mu_{\text{cl}}} \\
+ \int dz d\bar{z} \left( b(\bar{\partial}c + c\partial\mu_{\text{cl}} - \mu_{\text{cl}}\partial c) + \bar{b}(\partial\bar{c} + \bar{c}\partial\bar{\mu}_{\text{cl}} - \bar{\mu}_{\text{cl}}\bar{\partial}\bar{c}) \right) \ . \tag{8.30}$$

The ghost numbers and the dimensions of all the fields and sources are displayed in Table 8.1.

|  | $s$ | $\mathbf{X}$ | $\mu$ | $c$ | $b$ | $L$ | $\mathbf{Y}$ |
|---|---|---|---|---|---|---|---|
| Ghost number | 1 | 0 | 0 | 1 | -1 | -2 | -1 |
| Dimension | 1 | 0 | 0 | 0 | 1 | 1 | 1 |

**Table 8.1.** Ghost numbers and dimensions.

In particular, the expression (8.30) can be seen as describing the propagation of the quantized fields $(\mathbf{X}, c, \bar{c}, b, \bar{b})$ in a nontrivial classical background metric whose components are parametrized by $\mu_{\text{cl}}$. Equations (8.27), (8.29) represent the final version of the Slavnov–Taylor identity and of the complete classical action and will be taken as the starting point for the analysis of the quantum aspects of the model.

However, before going any further, let us spend some words on the physical meaning of the Beltrami differential (by now a *classical* variable). If on the one hand the introduction of the variable $\mu_{\text{cl}}$ allows to express in a natural way the dependence of the classical string action (8.1) on a conformal class of metrics, on the other hand it turns out to be very useful since it is directly related to the energy–momentum tensor, the components of the latter being given by [145, 146, 147]:

$$T = T_{zz} = -\frac{\delta S}{\delta \mu_{\text{cl}}} \ , \qquad \bar{T} = T_{\bar{z}\bar{z}} = -\frac{\delta S}{\delta \bar{\mu}_{\text{cl}}} \ . \tag{8.31}$$

It becomes apparent then that the Slavnov–Taylor identity (8.27) is the starting point for a field theory characterization of the string Green's functions with insertion of the energy–momentum tensor.

The explicit expression for the component $T$ is easily computed and reads:

$$T = T^{\mathbf{X}} + T^{\text{ghost}} + T^{\text{ext}} , \qquad (8.32)$$

with

$$T^{\mathbf{X}} = \mathcal{D}\mathbf{X} \cdot \mathcal{D}\mathbf{X} ,$$

$$T^{\text{ghost}} = (\partial b)c + 2b\partial c ,$$

$$T^{\text{ext}} = -\frac{(\bar{c} - \bar{\mu}_{\text{cl}}c)}{1 - \mu_{\text{cl}}\bar{\mu}_{\text{cl}}} \mathbf{Y} \cdot \mathcal{D}\mathbf{X} , \qquad (8.33)$$

where $\mathcal{D}$ denotes the *covariant* derivative (8.9) taken at $\mu = \mu_{\text{cl}}$:

$$\mathcal{D} = \frac{1}{1 - \mu_{\text{cl}}\bar{\mu}_{\text{cl}}}(\partial - \bar{\mu}_{\text{cl}}\bar{\partial}) . \qquad (8.34)$$

Let us also introduce for further use the linearized Slavnov–Taylor operator $\mathcal{S}_S$ corresponding to the classical action (8.30):

$$\mathcal{S}_S = \int dz d\bar{z} \left( \frac{\delta S}{\delta \mathbf{Y}} \cdot \frac{\delta}{\delta \mathbf{X}} + \frac{\delta S}{\delta \mathbf{X}} \cdot \frac{\delta}{\delta \mathbf{Y}} + \frac{\delta S}{\delta \mu_{\text{cl}}} \frac{\delta}{\delta b} + \frac{\delta S}{\delta b} \frac{\delta}{\delta \mu_{\text{cl}}} + \frac{\delta S}{\delta L} \frac{\delta}{\delta c} \right.$$
$$\left. + \frac{\delta S}{\delta c} \frac{\delta}{\delta L} + \frac{\delta S}{\delta \bar{\mu}_{\text{cl}}} \frac{\delta}{\delta \bar{b}} + \frac{\delta S}{\delta \bar{b}} \frac{\delta}{\delta \bar{\mu}_{\text{cl}}} + \frac{\delta S}{\delta \bar{L}} \frac{\delta}{\delta \bar{c}} + \frac{\delta S}{\delta \bar{c}} \frac{\delta}{\delta \bar{L}} \right) , \qquad (8.35)$$

which, as a consequence of (8.27), turns out to be nilpotent

$$\mathcal{S}_S \mathcal{S}_S = 0 . \qquad (8.36)$$

## 8.2 The Supersymmetry Algebra

In this section we show that, in analogy with the topological field theories discussed in Chap. 7, also the bosonic string theory possesses a supersymmetry algebra [153] whose generators are identified with the linearized Slavnov–Taylor operator (8.35) and with a pair of Ward operators carrying respectively a $z$ and a $\bar{z}$ index. In what follows, in order to avoid the technical complexities of working on a curved space (see e.g. [154] for the case of the Chern–Simons theory), we identify for simplicity the string world sheet $\mathcal{M}$ with the flat complex plane $\mathbb{C}$. The result may be consistently adapted to a general two–dimensional manifold by introducing an appropriate projective connection (see [147]).

*Remark.* One should note that in the flat limit the Beltrami parameters $(\mu_{\text{cl}}, \bar{\mu}_{\text{cl}})$ have no more the meaning of a background metric. Instead, they are introduced as purely external fields [146] (see also [155, 156] for the higher dimensional case) needed in order to properly define the components $(T, \overline{T})$ of the energy–momentum tensor (8.33) which, being quadratic in the fields, are composite operators. As any external field (cf. Sect 2.1) $(\mu_{\text{cl}}, \bar{\mu}_{\text{cl}})$ have to be set to zero at the end of the computation (this would be not true for a curved space–time, in this case $(\mu_{\text{cl}}, \bar{\mu}_{\text{cl}})$ identifying the metric). Let us also point out that in the flat limit the classical action (8.30) loses its interpretation in terms of string theory. However as shown in [146] it can be viewed as a useful simple example of a two dimensional conformal invariant theory. The Slavnov–Taylor identity (8.27) describes thus the conformal properties of the stress–energy tensor of a free theory in flat space–time. In particular, a $\mu_{cl}$–dependent anomaly in the quantum extension of (8.27) will have the meaning of a central charge in the energy–momentum current algebra [53].

In order to show the existence of such a supersymmetric structure let us introduce the two integrated functional operators $(\mathcal{W}, \overline{\mathcal{W}})$:

$$\mathcal{W} = \int dz d\bar{z} \left( \bar{\mu}_{cl} \frac{\delta}{\delta \bar{c}} + \frac{\delta}{\delta c} + \bar{L} \frac{\delta}{\delta \bar{b}} \right) , \tag{8.37}$$

and

$$\overline{\mathcal{W}} = \int dz d\bar{z} \left( \mu_{cl} \frac{\delta}{\delta c} + \frac{\delta}{\delta \bar{c}} + L \frac{\delta}{\delta b} \right) , \tag{8.38}$$

which, together with the linearized Slavnov–Taylor operator $\mathcal{S}_S$, are easily seen to obey the following anticommutation relations

$$\{ \mathcal{S}_S, \mathcal{W} \} = \mathcal{P} , \qquad \{ \mathcal{S}_S, \overline{\mathcal{W}} \} = \overline{\mathcal{P}} , \tag{8.39}$$

$$\{ \mathcal{W}, \mathcal{W} \} = \{ \mathcal{W}, \overline{\mathcal{W}} \} = \{ \overline{\mathcal{W}}, \overline{\mathcal{W}} \} = 0 , \tag{8.40}$$

where $(\mathcal{P}, \overline{\mathcal{P}})$ denote the translation Ward operators

$$\mathcal{P} = \sum_{\text{all fields}} \int dz d\bar{z} \, \partial\varphi \frac{\delta}{\delta\varphi} , \qquad \overline{\mathcal{P}} = \sum_{\text{all fields}} \int dz d\bar{z} \, \bar{\partial}\varphi \frac{\delta}{\delta\varphi} . \tag{8.41}$$

One sees from (8.39) that the algebra between $(\mathcal{W}, \overline{\mathcal{W}})$ and $\mathcal{S}_S$ closes on the translations, thus exhibiting a supersymmetric structure of the Wess–Zumino type.

In complete analogy with the topological models of Chap. 7, one then has the linearly broken Ward identities:

$$\mathcal{W}S = \Delta_{cl} , \qquad \overline{\mathcal{W}}S - \overline{\Delta}_{cl} , \tag{8.42}$$

with $(\Delta_{cl}, \overline{\Delta}_{cl})$ given by

$$\Delta_{cl} = \int dz d\bar{z} \left( \bar{L}\partial\bar{c} + L\partial c - \bar{b}\partial\bar{\mu}_{cl} - b\partial\mu_{cl} - \mathbf{Y} \cdot \partial\mathbf{X} \right) , \tag{8.43}$$

and

$$\overline{\Delta}_{cl} = \int dz d\bar{z} \left( L\bar{\partial}c + \bar{L}\bar{\partial}\bar{c} - b\bar{\partial}\mu_{cl} - \bar{b}\bar{\partial}\bar{\mu}_{cl} - \mathbf{Y} \cdot \bar{\partial}\mathbf{X} \right) . \tag{8.44}$$

The breakings (8.43), (8.44), being linear in the quantum fields, are classical, i.e., not subjected to renormalization.

Let us point out that, on a more general level, the Ward operators $(\mathcal{W}, \overline{\mathcal{W}})$, independently from the fact that they are associated or not with a (possibly linearly broken) symmetry of the classical action, have an intrinsic meaning. Indeed, from (8.39), (8.40), they can be seen as algebraic abstract operators which allow to decompose the space–time translations $(\mathcal{P}, \overline{\mathcal{P}})$ as BRS anticommutators. This property, as it has been shown in Sect. 5.4 and in Chap. 7, can be used as a powerful and elegant tool for solving the descent equations associated with the integrated $\mathcal{S}_S$-cohomology. It gives then an algebraic characterization of the anomalies and of the invariant counterterms which can arise at the quantum level [116, 142]. This will be the task of the next section.

## 8.3 The Diffeomorphism Anomaly and the Critical Dimension

We face now the problem of extending the classical Slavnov–Taylor identity (8.27) to the quantum level.

The absence of interaction terms between the propagating fields $(\mathbf{X}, c, \bar{c}, b, \bar{b})$ in the action (8.30), implies that the whole set of one–particle irreducible Feynman diagrams reduces to one–loop amplitudes with an arbitrary number of external legs $(\mu_{cl}, \bar{\mu}_{cl})$. Hence

$$\Gamma = S + \hbar \Gamma^{(1)} ,\tag{8.45}$$

where $\Gamma^{(1)}$ represents the one–loop contributions to the vertex functional.

The classical Slavnov–Taylor identity (8.27) extends then to the quantum level as

$$\mathcal{S}(\Gamma) = \hbar k \mathcal{A} ,\tag{8.46}$$

where $k$ is a one–loop numerical factor and $\mathcal{A}$ is an integrated local two–form of ghost number one and of dimension three[2]:

$$\mathcal{A} = \int \mathcal{A}_2^1 ,\tag{8.47}$$

which obeys the consistency condition

$$\mathcal{S}_S \mathcal{A} = 0 .\tag{8.48}$$

As in the case of the gauge theories (see Sect. 4.3), the consistency condition (8.48) yields, at the level of the integrand, a tower of descent equations:

$$\mathcal{S}_S \mathcal{A}_2^1 + d\mathcal{A}_1^2 = 0 ,$$
$$\mathcal{S}_S \mathcal{A}_1^2 + d\mathcal{A}_0^3 = 0 ,\tag{8.49}$$
$$\mathcal{S}_S \mathcal{A}_0^3 = 0 ,$$

where $d$ denotes the exterior derivative

$$d = dz\, \partial + d\bar{z}\, \bar{\partial} ,\tag{8.50}$$

with

$$d^2 = 0 \quad , \quad \{\mathcal{S}_S , d\} = 0 .\tag{8.51}$$

The Ward operators (8.37), (8.38) allow [153] to find a particular solution of the ladder equations (8.49) from the knowledge of the nontrivial solution of the last of these equations (which is a problem of local cohomology instead of a modulo–$d$ one). It is easy to check indeed that, once $\mathcal{A}_0^3$ is known, the remaining cocycles $\mathcal{A}_1^2$ and $\mathcal{A}_2^1$ are identified with the $(\mathcal{W}, \overline{\mathcal{W}})$–transform of $\mathcal{A}_0^3$, i.e., (cf. Sect 5.4)

$$\mathcal{A}_1^2 = (\mathcal{W}\mathcal{A}_0^3)\, dz + (\overline{\mathcal{W}}\mathcal{A}_0^3)\, d\bar{z} ,$$
$$\mathcal{A}_2^1 = (\mathcal{W}\overline{\mathcal{W}}\mathcal{A}_0^3)\, dz \wedge d\bar{z} .\tag{8.52}$$

---

[2]We adopt here the usual convention of denoting with $\mathcal{A}_q^p$ a $q$–form with ghost number $p$.

*Remark.* This is quite analog to what happens in the topological theories studied in Chap. 7.

It turns out that, due to the vanishing of the local cohomology of $\mathcal{S}_S$ in the sector of one–forms with ghost number two and in the sector of two–forms with ghost number one [152, 153], the expression (8.52) is in fact the most general solution of the descent equations (8.49), modulo trivial cocycles.

For what concerns the local cohomology of $\mathcal{S}_S$ in the zero–form sector with ghost number three it turns out [152, 153] that the general nontrivial solution for $\mathcal{A}_0^3$ contains two elements:

$$\mathcal{A}_0^3 = \left( \Omega^{(1)}, \ \Omega^{(2)} \right) , \tag{8.53}$$

which, modulo a $\mathcal{S}_S$–coboundary, can be written as

$$\Omega^{(1)} = \left( c\partial c\partial^2 c - \bar{c}\bar{\partial}\bar{c}\bar{\partial}^2\bar{c} \right) , \tag{8.54}$$

and

$$\Omega^{(2)} = \left( c\bar{c}\partial c + c\bar{c}\bar{\partial}\bar{c} \right) \mathcal{D}\mathbf{X} \cdot \overline{\mathcal{D}}\mathbf{X} f(\mathbf{X}) , \tag{8.55}$$

$f(\mathbf{X})$ denoting a real arbitrary formal power series in the matter fields $\mathbf{X}$ without constant term:

$$f(\mathbf{X}) = \sum_{n=1}^{\infty} f_n (\mathbf{X} \cdot \mathbf{X})^n . \tag{8.56}$$

We remark however that to the cocycle $\Omega^{(2)}$ one cannot associate a true anomaly. Indeed from the absence in the classical action (8.29) of self–interaction terms in the matter fields it is easily seen that, in spite of the fact that $\Omega^{(2)}$ is cohomologically nontrivial, the numerical coefficient of the corresponding Feynman diagrams automatically vanishes. We turn then to the analysis of the expression (8.54).

Using (8.52) for the cocycle $\mathcal{A}_2^1$ corresponding to $\Omega^{(1)}$ one gets the expression

$$\mathcal{A}_2^1 = \left( -\partial\mu_{\mathrm{cl}}\partial^2 c + \partial c\partial^2\mu_{\mathrm{cl}} - \bar{\partial}\bar{\mu}_{\mathrm{cl}}\bar{\partial}^2\bar{c} + \bar{\partial}\bar{c}\bar{\partial}^2\bar{\mu}_{\mathrm{cl}} \right) dz \wedge d\bar{z} . \tag{8.57}$$

Finally, after integration, the anomalous Slavnov–Taylor identity (8.46) takes the form

$$\mathcal{S}(\Gamma) = 2\hbar k \int d^2 z \left( \mu_{\mathrm{cl}}\partial^3 c + \bar{\mu}_{\mathrm{cl}}\bar{\partial}^3\bar{c} \right) . \tag{8.58}$$

Let us remark that the anomaly splits in two terms, depending on either one of the two pairs $(\mu_{\mathrm{cl}}, c)$ or $(\bar{\mu}_{\mathrm{cl}}, \bar{c})$. It thus respects the factorization property already observed in the BRS transformations (8.16).

In particular, it follows from this expression that the numerical coefficient $k$ is related to the one–loop Feynman diagrams with two $\mu_{\mathrm{cl}}$ (or $\bar{\mu}_{\mathrm{cl}}$) external legs, as one can easily check by acting on (8.58) with the functional derivative

$$\frac{\delta}{\delta c(u)} \frac{\delta}{\delta \mu_{\mathrm{cl}}(w)} , \tag{8.59}$$

and then setting all the fields and sources equal to zero.

The explicit value of $k$ turns out [146] to be proportional to $(D - 26)$. This implies that the model can be consistently quantized only if the dimension of the Euclidean target space is equal to the critical value of 26.

Let us conclude this section by remarking that the diffeomorphism anomaly
(8.58) is completely equivalent [157] to the original string Weyl anomaly found by
Polyakov [150]. It is well known indeed that, using a nonpolynomial local Wess–
Zumino action [107, 158] as a counterterm, one can shift a diffeomorphism anomaly
into a Weyl anomaly and vice versa. This Wess–Zumino action (called in this case the
Liouville action $\Gamma_{\text{Liouv}}$) depends on $(\mu_{\text{cl}}, \bar{\mu}_{\text{cl}})$ and on an *additional field* $\rho$ transforming
under BRS in such a way that

$$s\Gamma_{\text{Liouv}} = -\hbar k \mathcal{A} \ .$$

This additional field is nothing else than the conformal factor appearing in the Beltrami parametrization (8.3) of the metric. One may then compensate the diffeomorphism anomaly (8.58) by adding the Liouville action as counterterm. However, as one
can expect from the explicit presence of the scaling factor $\rho$, the resulting diffeomorphism invariant theory turns out to be anomalous under Weyl transformations.

## 8.4 The Algebra of the Energy–Momentum Tensor

As we have already said in the previous sections, the presence of a $\mu_{\text{cl}}$–dependent
anomaly in the quantum extension of the Slavnov–Taylor identity has the meaning of
a central charge in the current algebra of the energy–momentum tensor.

For a better understanding of this point let us discuss in detail the relation between
the coefficient $k$ of the anomalous term (8.58) and the central charge $\lambda$ of the two-
dimensional stress–energy
algebra. The latter, as it is well known [148, 159], can be identified through the
behaviour of the Green function with insertion of two energy–momentum tensors, i.e.,

$$\langle T(z)T(z')\rangle = \frac{\hbar\lambda}{2}\frac{1}{(z-z')^4} \ . \tag{8.60}$$

Making use of the Legendre transformation (2.17)

$$\Gamma = Z^{\text{c}}(\mu_{\text{cl}}, \bar{\mu}_{\text{cl}}, L, \bar{L}, \mathbf{Y}, J) - \int d^2z \left( \mathbf{J}_X \cdot \mathbf{X} + J_c c + J_{\bar{c}}\bar{c} + J_b b + J_{\bar{b}}\bar{b} \right) \tag{8.61}$$

$$\varphi = \frac{\delta Z^{\text{c}}}{\delta J} \ , \qquad \varphi = (\mathbf{X}, c, \bar{c}, b, \bar{b}) \ ,$$

the anomalous Slavnov–Taylor identity (8.58) can be rewritten in terms of the connected functional $Z^{\text{c}}$ as

$$\int d^2z \left( -\mathbf{J}_X \cdot \frac{\delta Z^{\text{c}}}{\delta \mathbf{Y}} + J_b \frac{\delta Z^{\text{c}}}{\delta \mu_{\text{cl}}} + J_c \frac{\delta Z^{\text{c}}}{\delta L} + J_{\bar{b}}\frac{\delta Z^{\text{c}}}{\delta \bar{\mu}_{\text{cl}}} + J_{\bar{c}}\frac{\delta Z^{\text{c}}}{\delta \bar{L}} \right)$$
$$= 2\hbar k \int d^2z \left( \mu_{\text{cl}}\partial^3 \frac{\delta Z^{\text{c}}}{\delta J_c} + \bar{\mu}_{\text{cl}}\bar{\partial}^3 \frac{\delta Z^{\text{c}}}{\delta J_{\bar{c}}} \right) \ . \tag{8.62}$$

Acting on both sides of (8.62) with the operator

$$\frac{\delta}{\delta J_b(z')}\frac{\delta}{\delta \mu_{\text{cl}}(z)} \ , \tag{8.63}$$

and setting all the sources equal to zero one gets the equation

$$\frac{\delta^2 Z^c}{\delta\mu_{cl}(z')\delta\mu_{cl}(z)}\bigg|_{J=\mu_{cl}=0} = 2\hbar k \partial_z^3 \frac{\delta^2 Z^c}{\delta J_b(z')\delta J_c(z)}\bigg|_{J=\mu_{cl}=0} \tag{8.64}$$

The term $\delta^2 Z^c/\delta J_b \delta J_c$ in the right–hand side of (8.64) is nothing but the propagator $\langle b(z')c(z)\rangle$ of the $(b-c)$ ghost system. The latter, as one can see from (8.30), is the Green function for the operator $\bar{\partial}$ and is given by

$$\langle b(z')c(z)\rangle = \frac{1}{\pi(z-z')} , \tag{8.65}$$

as a distribution on the plane, due to the well known relation [159]

$$\bar{\partial}_z \frac{1}{(z-z')} = \pi\delta^2(z-z') . \tag{8.66}$$

Recalling now that $\mu_{cl}(z)$ is the external source for the component $T$ of the energy–momentum tensor (see (8.31)), equation (8.64) implies

$$\langle T(z)T(z')\rangle = \frac{2\hbar k}{\pi}\partial_z^3 \frac{1}{(z-z')} = -\frac{12\hbar k}{\pi}\frac{1}{(z-z')^4} , \tag{8.67}$$

which clearly shows that the anomalous Slavnov–Taylor identity (8.58) provides an algebraic characterization of the energy–momentum current algebra. The classical action (8.30) describes thus a model whose corresponding central charge takes the value $\lambda = -24k/\pi$.

# References

[1] "Selected Papers on Quantum Electrodynamics" *(ed. J. Schwinger) Dover, New York, 1958*;

[2] R.E. Marshak, "Conceptual Foundations of Modern Particle Physics", *World Scientific, Singapore, 1993*;

[3] A. Ashtekar, "Non–Perturbative Canonical Gravity", *World Scientific, Singapore, 1991*;
A. Ashtekar, "Mathematical Problems of Non–Perturbative Quantum General Relativity", *Lectures delivered at the 1992 Les Houches summer school on Gravitation and Quantization (gr-qc/9302024)*;

[4] J.R. Oppenheimer, *Phys. Rev.* 35 (1930) 461;

[5] N.N. Bogoliubov and D.V. Shirkov, "Introduction to the Theory of Quantized Fields", *J. Wiley, New York, 1980 (third edition)*;

[6] H. Epstein and V. Glaser, *Ann. Inst. Henri Poincaré* 19 (1973) 211;

[7] "Renormalization Theory" *(eds. G. Velo and A.S. Wightman) D. Reidel, Dordrecht, 1976*;

[8] G. 't Hooft, "Gauge Theory and Renormalization", *Talk given at the International Conference on: "The History of Original Ideas and Basic Discoveries in Particle Physics", Erice, 1994 (hep-th/9410038)*;

[9] R.P. Feynman, "Quantum Electrodynamics", *W. A. Benjamin, New York 1962*;
J.D. Bjorken and S.D. Drell, "Relativistic Quantum Fields", *McGraw-Hill, New York, 1965*;

[10] G. 't Hooft, *Nucl. Phys.* B33 (1971) 173; *Nucl. Phys.* B35 (1971) 167;

[11] G. 't Hooft and M. Veltman, *Nucl. Phys.* B50 (1972) 318;

[12] J.H. Lowenstein, *Phys. Rev.* D4 (1971) 2281;
J.H. Lowenstein, *Commun. Math. Phys.* 24 (1971) 1;

[13] Y.M.P. Lam, *Phys. Rev.* D6 (1972) 2145; *Phys. Rev.* D7 (1973) 2943;

[14] T.E. Clark and J.H. Lowenstein, *Nucl. Phys.* B113 (1976) 109;

[15] C. Becchi, A. Rouet and R. Stora, *Phys. Lett.* B52(1974)344, *Commun. Math. Phys.* 42 (1975) 127;

[16] C. Becchi, A. Rouet and R. Stora, *Ann. Phys. (N.Y.)* 98 (1976) 287;

[17] C. Becchi, A. Rouet and R. Stora, "Renormalizable theories with symmetry breaking", *in: "Field Theory, Quantization and Statistical Physics" (ed. E. Tirapegui) D. Reidel, Dordrecht, 1981*;
and "Renormalizable models with broken symmetries", *in: "Renormalization Theory", p.299 (eds. G. Velo and A.S. Wightman) D. Reidel, Dordrecht, 1976*;

[18] J. Glimm and A. Jaffe, "Quantum Physics: a Functional Point of View", *Springer Verlag, Berlin, 1981*;

[19] T. Kinoshita, "Quantum Electrodynamics", *World Scientific, Singapore, 1990*;

[20]C. Rubbia, *Nuovo Cimento* 107A (1994) 1001;

G. Altarelli, "Status of precision tests of electroweak theory", *Conference on Phenomenology of Unification: from Present to Future, Rome, Italy, 1994, Preprint CERN-TH-7319/94*;

[21]"QCD: 20 Years Later" *(eds. P.M. Zervas and H.A. Kastrup) World Scientific, Singapore, 1993*;

[22]M. Henneaux and C. Teitelboim, "Quantization of Gauge systems", *Princeton Univ. Press, 1992*;

[23]D.M. Gitman and I.V. Tyutin, "Quantization of Fields with Constraints", *Springer Verlag, Berlin, 1990*;

[24]W. Zimmermann, *Commun. Math. Phys.* 15 (1969) 208;

[25]J.H. Lowenstein, *Commun. Math. Phys.* 47 (1976) 53;

[26]O. Piguet and K. Sibold, "Renormalized Supersymmetry", *series "Progress in Physics", vol. 12 (Birkhäuser Boston Inc., 1986)*;

[27]R. Haag, "Local Quantum Physics", *Springer Verlag, Berlin, 1992*;

[28]K. Osterwalder and R. Schrader, *Commun. Math. Phys.* 31 (1973) 83;
*Commun. Math. Phys.* 42 (1975) 281; *Helv. Phys. Acta* 46 (1973) 272;

[29]E.C.G.Stueckelberg and D. Rivier, *Helv. Phys. Acta* 23 (1950) 215;

[30]R.P. Feynman, *Phys. Rev.* 76 (1949) 749,769;

[31]P. Ramond, "Field Theory: a Modern Primer", *Frontiers in Physics, Benjamin/Cummings Publ. Comp. (1981)*;

[32]K. Symanzik, "Renormalization of theories with broken symmetry", *in: "Cargèse Lectures in Physic 1970", p.179 (ed. D. Bessis) Gordon & Breach, NY 1972*;

[33]S. Weinberg, *Phys. Rev.* 18 (1960) 838;

[34]W. Zimmermann, *Commun. Math. Phys.* 11 (1968) 1;

[35]C. Bollini and J. J. Giambiagi, *Phys. Lett.* B40 (1972) 566;
G. M. Cicuta and E. Montaldi, *Nuovo Cimento Lett.* 4 (1972) 329;
G. 't Hooft and M. Veltman, *Nucl. Phys.* B44 (1972) 189;

[36]P. Breitenlohner and D. Maison, "Some results on dimensional regularization", *in: "Renormalization Theory" (eds. G. Velo and A.S. Wightman) D. Reidel, Dordrecht, 1976*;
P. Breitenlohner and D. Maison, *Commun. Math. Phys.* 52 (1977) 11,39,55;

[37]E. Speer, "Dimensional and analytic renormalization", *in: "Renormalization Theory" (eds. G. Velo and A.S. Wightman) D. Reidel, Dordrecht, 1976*;

[38]J. Polchinski, *Nucl. Phys.* B231 (1984) 269;

[39]G. Keller and C. Kopper, *Phys. Lett.* B273 (1991) 323;
*Commun. Math. Phys.* 148 (1992) 445;

[40]C. Becchi, "On the Construction of Renormalized Quantum Field Theory Using Renormalization Group Techniques", *in: "Elementary Particles, Quantum Fields and Statistical Mechanics" (eds. M. Bonini, G. Marchesini and E. Onofri) Parma University, Italy, 1993*;

[41]J. Zinn–Justin, "Quantum Field Theory and Critical Phenomena", *Oxford Science Publications, Clarendon Press, Oxford, second edition, 1993*;

[42]C. Itzykson and J.-B. Zuber, "Quantum field theory", *McGraw–Hill 1985*;

[43]J. Collins, "Renormalization", *Cambridge University Press 1984*;

[44]W. Zimmermann, "Remark on equivalent formulations for Bogoliubov's method of renormalization", *in: "Renormalization Theory" (eds. G. Velo and A.S. Wightman) D. Reidel, Dordrecht, 1976*;

[45]W. Zimmermann, *Ann. Phys. (N.Y.)* 77 (1973) 536;

[46]J.H. Lowenstein, "Seminars on Renormalization Theory vol. II", *University of Maryland Techn. Rep. No 73-068 (1972)*;

[47]J.H. Lowenstein, "BPHZ renormalization", *in: "Renormalization Theory" (eds. G. Velo and A.S. Wightman) D. Reidel, Dordrecht, 1976*;

[48]G. Popineau and R. Stora, "A pedagogical remark on the main theorem of perturbative renormalization theory", *(Thèse de Troisième Cycle) unpublished*;

[49]O. Steinmann, *Commun. Math. Phys.* 152 (1993) 627;

[50]R. Oehme and W. Zimmermann, *Commun. Math. Phys.* 97 (1985) 569;
R. Oehme, K. Sibold and W. Zimmermann, *Phys. Lett.* B147 (1984) 115;
W. Zimmermann, *Commun. Math. Phys.* 97 (1985) 211;
R. Oehme, *Progr. theor. Phys. suppl.* 86 (1986) 215

[51]R. Oehme, K. Sibold and W. Zimmermann, *Phys. Lett.* B153 (1985) 142;

[52]C. Becchi, A. Blasi, G. Bonneau, R. Collina and F. Delduc, *Commun. Math. Phys.* 120 (1988) 121;

[53]C. Becchi and O. Piguet, *Nucl. Phys.* B315 (1989) 153; *Nucl. Phys.* B347 (1990) 596;

[54]G. Bonneau, *Phys. Lett.* 333 (1994) 46;

[55]K. Symanzik, "Small distance behaviour in field theory", in *Springer Tracts in Modern Physics 57 (1971) 222*;

[56]J. Schwinger, *Phys. Rev.* 82 (1951) 918;
C.S. Lam, *Nuovo Cimento* 38 (1965) 1754;

[57]C.G. Callan, *Phys. Rev.* D2 (1970) 1541;
K. Symanzik, *Commun. Math. Phys.* 18 (1970) 227;

[58]A. Blasi and R. Collina, *Nucl. Phys.* B285 (1987) 204; *Phys. Lett.* B200 (1988) 98;

[59]I.V. Tyutin, "Gauge invariance in field theory and statistical mechanics", *Lebedev preprint FIAN, n⁰ 39 (1975), unpublished*;

[60]B. Zumino, "Anomalies, Cocycles and Schwinger Terms", *in: "Symposium on Anomalies, Geometry, Topology", (ed. W.A. Bardeen and A.R. White) World Scientific, Singapore, 1985*;

[61]O. Piguet, "Renormalisation en théorie quantique des champs" and "Renormalisation des théories de jauge", lectures of the *"Troisième cycle de la physique en Suisse Romande" (1982–1983)*;

[62]J.H. Lowenstein, *Commun. Math. Phys.* 24 (1971) 1;

[63]C. Becchi, "An Introduction to the Renormalization of Theories with Continuous Symmetries, to the Chiral Models and to their Anomalies", *in: "Fields and Particles" (ed. H.Mitter and W. Schweiger) Springer–Verlag, Berlin, Heidelberg, 1990*;

[64]V.S. Varadarajan, "Lie groups, Lie Algebras and their Representations", *Springer–Verlag, Berlin, 1984*;

[65]T.Kugo and I. Ojima, *Phys. Lett.* B73 (1978) 459;
*Progr. Theor. Phys. Suppl.* 66 (1979) 1;

[66]C. Becchi, "Lectures on the Renormalization of Gauge Theories", *in: Relativity, Groups and Topology II, Les Houches 1983 (ed. B.S. DeWitt and R. Stora) North Holland, Amsterdam, 1984*;

[67]B. Lautrup, *Mat. Fys. Medd. Dan. Vid. Selsk* 35 (1967), No.11;
N. Nakanishi, *Progr. Theor. Phys.* 35 (1966) 1111; *Progr. Theor. Phys.* 37 (1967) 618;

[68]J. Zinn-Justin, "Renormalization of Gauge Theories", *in: "Trends in Elementary Particle Theory", Lecture Notes in Physics 37, Springer Verlag, 1975*;

[69]O. Piguet and A. Rouet, *Phys. Rep.* 76 (1981) 1;

[70]"Physical and Nonstandard Gauges" *(eds. P. Gaigg, W. Kummer and M. Schweda) Lectures Notes in Physics, Vol.361 Springer–Verlag, Berlin, Heidelberg, 1990*;

[71] A. Bassetto, G. Nardelli and R. Soldati, "Yang–Mills Theories in Algebraic Non–Covariant Gauges", *World Scientific, 1991*;

[72] W. Kummer, *Acta Physica Austriaca, Suppl.* XV (1976) 423;

[73] S. Mandelstam, *Nucl. Phys.* B213 (1983) 149;
G. Leibbrandt, *Phys. Rev.* D29 (1984) 1699; *Rev. Mod. Phys.* 59 (1987) 1067;

[74] J. Zinn–Justin, *Nucl. Phys.* B246 (1984) 246;

[75] E. Kraus and K. Sibold, *Nucl. Phys.* B331 (1990) 350;

[76] F. Delduc and S.P.Sorella, *Phys. Lett.* B231 (1989) 408;

[77] A. Rouet, *Commun. Math. Phys.* 48 (1976) 89;

[78] G. Barnich and M.Henneaux, *Phys. Rev. Lett.* 72 (1994) 1588;

[79] G. Barnich, F. Brandt and M.Henneaux, "Local BRST cohomology in the antifield formalism I and II", *Preprints ULB-TH-94/06,07 (hep-th/9405109 and 9405194)*;

[80] G. Bandelloni, C. Becchi, A. Blasi and R. Collina,
*Ann. Inst. Henri Poincaré* 28 (1978) 225,255;

[81] F. Brandt, *Phys. Lett.* B320 (1994) 57;

[82] J. Wess and B. Zumino, *Phys. Lett.* B49 (1974) 52;

[83] J. Manes, R. Stora and B. Zumino, *Commun. Math. Phys.* 102 (1985) 157;

[84] B. Zumino, *Nucl. Phys.* B253 (1985) 477;

[85] S. L. Adler, *Phys. Rev.* 117 (1969) 2426;
J. S. Bell and R. Jackiw, *Nuovo Cimento* 60 (1969) 47;
J. Schwinger, *Phys. Rev.* 82 (1951) 664;

[86] W. A. Bardeen, *Phys. Rev.* 184 (1969) 48;

[87] G. Costa, J. Julve, T. Marinucci and M. Tonin, *Nuovo Cimento* 38A (1977) 373;

[88] G. Bandelloni, C. Becchi, A. Blasi and R. Collina,
*Commun. Math. Phys.* 72 (1980) 239;

[89] O. Piguet and S.P. Sorella, *Nucl. Phys.* B381 (1992) 373;

[90] O. Piguet and S.P. Sorella, *Nucl. Phys.* B395 (1993) 661;

[91] M. Böhm, A. Denner, F. Jegerlehner, E. Kraus and K. Sibold, *book in preparation*;

[92] H. Kluberg-Stern and J.B. Zuber, *Phys. Rev.* D12 (1975) 467,482,3159;

[93] O. Piguet and K. Sibold, *Nucl. Phys.* B253 (1985) 517;

[94] J. Dixon, "Cohomology and renormalization of gauge theories", I, II, III, *unpublished preprints, 1976–1977*;
J. Dixon, *Commun. Math. Phys.* 139 (1991) 495;

[95] M. Dubois-Violette, M. Henneaux, M. Talon and C.M. Viallet,
*Phys. Lett.* B289 (1992) 361;

[96] F. Brandt, N. Dragon and M. Kreuzer, *Phys. Lett.* B231 (1989) 263;
*Nucl. Phys.* B332 (1990) 224, 250;
F. Brandt, PhD Thesis (in German), *University of Hanover (1991), unpublished*;

[97] G. Barnich, "Local BRST Cohomology in Yang-Mills Theory", PhD Thesis, *Université Libre de Bruxelles (1995)*;

[98] S. Ouvry, R. Stora and P. Van Baal, *Phys. Lett.* B220 (1989) 159;
A. Blasi and R. Collina, *Phys. Lett.* B222 (1989) 419;
S. P. Sorella, *Phys. Lett.* B228 (1989) 159;

[99] A. Brandhuber, O. Moritsch, M.W. de Oliveira, O. Piguet and M. Schweda, "A Renormalized Supersymmetry in the Topological Yang–Mills Field Theory", *preprint REF. TUW 94-10 – UGVA-DPT 1994/07-858 (hep-th/9407105)*;

[100] G. de Rham, "Variétés différentiables", *Hermann, Paris, 1973*;
and "Differentiable Manifolds" (English translation), *Springer–Verlag, Berlin, Heidelberg, 1984*;

[101]R. Stora, "Continuum Gauge Theories", *in: "New Developments in Quantum field Theory and Statistical Physics" (eds. M. Levy and P. Mitter) NATO ASI Series B26, Plenum, 1977;*

[102]C. Lucchesi, O. Piguet and K. Sibold, *Int. J. Mod. Phys.* A2 (1987) 385;

[103]C. Becchi, *Commun. Math. Phys.* 39 (1975) 329;

[104]J. Thierry–Mieg, University of Paris–Sud Thesis (1978); *J. Math. Phys.* 21 (1980) 2834; *Nuovo Cimento* 56A (1980) 396;

[105]B. Zumino, Wu Y.S., A. Zee, *Nucl. Phys.* B239 (1984) 477;

[106]B. Zumino, "Chiral Anomalies and Differential Geometry", *in: "Relativity, Groups and Topology II", Les Houches, 1983 (ed. B.S. DeWitt and R. Stora) North Holland, Amsterdam, 1984;*

[107]W. A. Bardeen and B. Zumino, *Nucl. Phys.* B244 (1984) 421;

[108]R.D. Ball, *Phys. Rep.* 182 (1989) 1;

[109]R. Stora, "Algebraic Structure and Topological Origin of Anomalies", *in: "Progress in Gauge Field Theory" (ed. 't Hooft et al.) Plenum Press, New York, London, 1984;*

[110]M. Dubois–Violette, M. Talon and C.M. Viallet,
*Commun. Math. Phys.* 102 (1985) 105;

[111]A.S. Schwarz, *Lett. Math. Phys.* 2 (1978) 247; *Commun. Math. Phys.* 67 (1979) 1;

[112]E. Witten, *Commun. Math. Phys.* 117 (1988) 353;
*Commun. Math. Phys.* 118 (1988) 411;

[113]D. Birmingham, M. Blau, M. Rakowski and G. Thompson, *Phys. Rep.* 209 (1991) 129;

[114]I.A. Batalin and G.A. Vilkovisky, *Phys. Lett.* B102 (1981) 27;
*Phys. Rev.* D28 (1983) 2567;

[115]M. Abud, J.-P. Ader and L. Cappiello, *Nuovo Cimento* 105A (1992) 1507;

[116]S.P. Sorella, *Commun. Math. Phys.* 157 (1993) 231;
S.P. Sorella and L. Tataru, *Phys. Lett.* B324 (1994) 351;

[117]O. Moritsch, M. Schweda and S.P. Sorella, *Class. Quantum Grav.* 11 (1994) 1225;

[118]S.L. Adler and W.A. Bardeen, *Phys. Rev.* 182 (1969) 1517;
W. A. Bardeen, "Renormalization of Yang–Mills fields and application to particle physics", *C.N.R.S. (Marseille) report 72/p470, 29 (1972);*
W. Bardeen, *proceedings of the 16th International Conference on High Energy Physics, Fermilab 1972, vol. II, p.295;*

[119]A. Blasi, O. Piguet and S. P. Sorella, *Nucl. Phys.* B356 (1991) 154;

[120]A. Zee, *Phys. Rev. Lett.* 29 (1972) 1198;

[121]J.H. Lowenstein and B. Schroer, *Phys. Rev.* D7 (1975) 1929;

[122]E. Witten, *Commun. Math. Phys.* 121 (1989) 351;

[123]A. Blasi and R. Collina, *Nucl. Phys.* B345 (1990) 472;

[124]F. Delduc, C. Lucchesi, O. Piguet and S. P. Sorella, *Nucl. Phys.* B346 (1990) 313;

[125]C. Nash and S. Sen,"Topology and Geometry for Physicists", *Academic Press, London, 1983;*

[126]D. Birmingham, M. Rakowski and G. Thompson, *Nucl. Phys.* B329 (1990) 83;

[127]D. Birmingham and M. Rakowski, *Mod. Phys. Lett.* A4 (1989) 1753;

[128]F. Delduc, F. Gieres and S. P. Sorella, *Phys. Lett.* B225 (1989) 367;

[129]N. Maggiore and S. P. Sorella, *Nucl. Phys.* B377 (1992) 236;
N. Maggiore and S. P. Sorella, *Int. J. Mod. Phys.* A8 (1993) 929;

[130]C. Lucchesi, O. Piguet and S. P. Sorella, *Nucl. Phys.* B395 (1993) 325;

[131]D. Birmingham and M. Rakowski, *Phys. Lett.* B269 (1991) 103;

[132]O. Piguet and K. Sibold, *Nucl. Phys.* B253 (1985) 269;
O. Piguet and S. P. Sorella, *Helv. Phys. Acta* 63 (1990) 683;

[133]W. Chen, G. W. Semenoff and Y. S. Wu, *Phys. Rev.* D44 (1991) 1625;

[134]L. V. Avdeev, G. V. Grogoryev and D. I. Kazakov, *Nucl. Phys.* B382 (1992) 561;

[135]A. Blasi, N. Maggiore and S. P. Sorella, *Phys. Lett.* B285 (1992) 54;

[136]A. N. Kapustin and P. I. Pronin, *Phys. Lett.* B318 (1993) 465;

[137]G.V. Dunne, R. Jackiw and C.A. Trugenberger, *Ann. Phys. (N.Y.)* 194 (1989) 197;

  G.V. Dunne and C. A. Trugenberger, *Ann. Phys. (N.Y.)* 204 (1990) 281;

[138]R. Floreanini and R. Percacci, *Phys. Lett.* B224 (1989) 291;

  G.V. Dunne and C. A. Trugenberger, *Phys. Lett.* B247 (1990) 81;

  *Phys. Lett.* B248 (1990) 305;

[139]K.S. Gupta and A. Stern, "4D Edge Currents from 5D Chern–Simons Theory", *preprint ISU-NP-94-12/UAHEP949 (hep-th/9410216)*;

[140]G. T. Horowitz, *Commun. Math. Phys.* 125 (1989) 417;

  G. T. Horowitz and M. Srednicki, *Commun. Math. Phys.* 130 (1990) 83;

  M. Blau and G. Thompson, *Ann. Phys. (N.Y.)* 205 (1991) 130;

[141]J.C. Wallet, *Phys. Lett.* B235 (1990) 71;

[142]E. Guadagnini, N. Maggiore and S.P. Sorella, *Phys. Lett.* B255 (1991) 65;

[143]A. Blasi and N. Maggiore *Class. Quantum Grav.* 10 (1993) 37;

[144]J. Scherk, *Rev. Mod. Phys.* 47 (1975) 123;

  M. Green, J. Schwarz and E. Witten, "Superstring theory", vol. I, II; *Cambridge University Press, 1987*

  E. D'Hoker and D.H. Phong, *Rev. Mod. Phys.* 60 (1988) 917;

[145]L. Baulieu, C. Becchi and R. Stora, *Phys. Lett.* B180 (1986) 55;

  L. Baulieu and M. Bellon, *Phys. Lett.* B196 (1987) 142;

[146]C. Becchi, *Nucl. Phys.* B304 (1988) 513;

[147]R. Stora, *in: "Non Perturbative Quantum Field Theory" (eds. 't Hooft and al.) Nato ASI series B, vol. 185, Plenum Press (88*;

  S. Lazzarini and R. Stora, "Ward Identities for Lagrangian Conformal Models", *in: "Knots, Topology and Quantum Field Theory", 13th John Hopkins Workshop (ed L. Lusanna) World Scientific 1989*;

[148]A. A. Belavin, A. M. Polyakov and A. B. Zamolodchikov, *Nucl. Phys.* 241 (1984) 333;

[149]M. Kaku, "Introduction to Superstrings", *Springer–Verlag, Berlin, 1988*;

[150]A. Polyakov, *Phys. Lett.* B103 (1981) 207;

[151]O. Lehto, "Univalent Functions and Teichmüller Space", *Springer–Verlag, Berlin, 1987*;

[152]G. Bandelloni and S. Lazzarini, *J. Math. Phys.* 34 (1993) 5413;

[153]M. Werneck de Oliveira, M. Schweda and S. P. Sorella, *Phys. Lett.* B315 (1993) 93;

[154]C. Lucchesi and O. Piguet, *Phys. Lett.* B271 (1991) 350; *Nucl. Phys.* B381 (1992) 281;

[155]G. Bandelloni, C. Becchi, A. Blasi and R. Collina, *Nucl. Phys.* B197 (1982) 347;

[156]E. Kraus and K. Sibold, *Nucl. Phys.* B372 (1992) 113; *Nucl. Phys.* B398 (1993) 125; *Ann. Phys. (N.Y.)* 219 (1992) 349;

[157]M. Knecht, S. Lazzarini and F. Thuillier, *Phys. Lett.* B251 (1990) 279;

  M. Knecht, S. Lazzarini and R. Stora, *Phys. Lett.* B262 (1991) 25;

[158]L. Bonora, P. Pasti and M. Tonin, *Phys. Lett.* B149 (1984) 346;

[159]A. Polyakov, "Gauge Fields and Strings", *Harwood Academic Publishers, 1987*;

# Springer-Verlag
# and the Environment

We at Springer-Verlag firmly believe that an international science publisher has a special obligation to the environment, and our corporate policies consistently reflect this conviction.

We also expect our business partners – paper mills, printers, packaging manufacturers, etc. – to commit themselves to using environmentally friendly materials and production processes.

The paper in this book is made from low- or no-chlorine pulp and is acid free, in conformance with international standards for paper permanency.

# Lecture Notes in Physics

For information about Vols. 1–409
please contact your bookseller or Springer-Verlag

# New Series m: Monographs